New Trends and Challenges in Open Data

Edited by Vijayalakshmi Kakulapati

Published in London, United Kingdom

New Trends and Challenges in Open Data
http://dx.doi.org/10.5772/intechopen.1000267
Edited by Vijayalakshmi Kakulapati

Contributors
Vijayalakshmi Kakulapati, Claus Rinner, Vibhatha Abeykoon, Geoffrey Charles Fox, Istvan Elek, Aliyu Ismail Ishaq, Abubakar Usman, Ahmad Abubakar Suleiman, Mahmod Othman, Hanita Daud, Rajalingam Sokkalingam, Uthumporn Panitanarak, Muhammad Azrin Ahmad, Lillian Alvares, Kira Tarapanoff

First published in London, United Kingdom, 2023 by IntechOpen
IntechOpen is the global imprint of INTECHOPEN LIMITED, registered in England and Wales, registration number: 11086078, 5 Princes Gate Court, London, SW7 2QJ, United Kingdom

British Library Cataloguing-in-Publication Data
A catalogue record for this book is available from the British Library

Additional hard and PDF copies can be obtained from orders@intechopen.com

New Trends and Challenges in Open Data
Edited by Vijayalakshmi Kakulapati
p. cm.
Print ISBN 978-1-83769-592-8
Online ISBN 978-1-83769-591-1
eBook (PDF) ISBN 978-1-83769-593-5

Meet the editor

Prof. Vijayalakshmi Kakulapati received a Ph.D. in Computer Science and Engineering from Jawaharlal Nehru Technological University (JNTU), Hyderabad, India. She is currently a professor in the Department of Information Technology, Sreenidhi Institute of Science and Technology, Hyderabad, India. She has 26 years of industry and teaching experience and is a member of various professional bodies, including the Institute of Electrical and Electronics Engineers (IEEE), Association for Computing Machinery (ACM), Computer Science Teachers Association (CSTA), Life member of Indian Society for Technical Education (LMISTE), Life member of Computer Society of India (LMCSI), International Association of Computer Science and Information Technology (IACSIT), Fellow member of Institution of Electronics and Telecommunication Engineers (FIETE), and more. She has more than 200 publications in national and international journals and conferences, 35 book chapters, and 6 books to her credit. She has received numerous awards, including Excellence in Research, Best Reviewer, appreciation awards, and more. Her areas of research include theoretical and practical information retrieval problems as well as machine learning applied to large-scale textual applications. Her research has focused on retrieval models, query/document representations, term weighting, term proximity models, and learning to rank (machine-learned ranking functions). She is also passionate about seeing research problems applied to real-world problems, especially those dealing with large, complex data sets. Along these lines, she is working with evaluating and designing novel search algorithms for Web search and summarization. Currently, Dr. Kakulapati is working with big data analytics, health informatics, the Internet of Things, deep learning, artificial intelligence, and data sciences.

Contents

Preface

The release of public data might enable a more open, collaborative, efficient, and productive examination, and use of public information, which increases its value when shared. Approximately 10 years ago, this concept sparked turmoil in a culture where the release of public records was unheard of. From this point on, other developments emerged that would come to define the development of the open data movement all over the globe. Data transparency and protection will be crucial as countries adjust to the digital era. The next problem is finding a fair balance between public privacy and the need for data while also increasing public faith in the administration's capacity to handle sensitive data with care. The ongoing research demonstrating the advantages of public access to public data should be expanded upon. The open data community has been successful in establishing open data as a foundational principle for improved administration and societal progress. However, the widespread institutional reforms required to significantly improve public administrations' data usage and sharing have not yet materialized from this trend. Thus, there is a growing trend toward a different strategy, one in which openness is planned and prioritized around reaching a specific goal while keeping openness by default as the long-term goal rather than exposing data in isolation and more or less at random.

This book contains two sections. "Trends and Challenges of Open Data" and "Case Studies". Each section contains three chapters.

Dr. Vijayalakshmi Kakulapati
Professor and Associate Dean (R&D),
Department of Information Technology,
Sreenidhi Institute of Science and Technology,
Yamnampet, Ghatkesar, Hyderabad, Telangana, India

Section 1

Trends and Challenges of Open Data

Chapter 1

Analysis of Trends and Challenges of Public Open Data in Health Care Industry Using Artificial Intelligence

Vijayalakshmi Kakulapati

Abstract

Understanding the public open data being gathered and analyzed is necessary before we can discuss health data analytics and its function in the healthcare industry. A significant quantity of health data is also being obtained, kept, and analyzed, in addition to data on the operations and procedures of the commercial side of the health-care industry. Any information about a patient's or a population's health is referred to as "health data." Medical professionals and administrators may find areas that need improvement or are in danger by using data from the health industry. With this knowledge, they may take steps to improve any areas where patient care is deficient and elevate the standard of care for all patients. Lab findings, vital sign recordings, prescription diaries, and computerized medical records all include enormous amounts of data. A change in the patient's health or the possibility of experiencing a major consequence may be detected by physicians and nurses using artificial intelligence (AI) techniques to spot data trends. Due to the complexity and expansion of data in the healthcare sector, AI will be employed there with greater frequency. Numerous types of AI are already being utilized by health insurance companies, medical organizations, and biological sciences enterprises. Solutions can be put into three main categories: operational tasks, patient engagement and participation, and medication and diagnosis recommendations. The health sector uses AI and data engineering to improve the processing and analysis of health data, compensation settlements, and other clinical records. The objective of this chapter is to learn about the capabilities of AI in using public open data as well as the trends and challenges in patient data.

Keywords: AI, health care, data, trends, challenges, patient, application, algorithms

1. Introduction

This chapter outlines the principles of open data, especially open health data, and examines how these concepts connect to the field of health care. The idea of open data has broad applications across many industries, and the literature that has been written about it emphasizes its significance. Governments from all over the globe are already developing regulations to improve data transparency by laying out in-depth guidelines for how to handle information about the general welfare. Making this data freely accessible to the public in forms that support a range of purposes is becoming more important.

All types of data that are made freely accessible to the public are collectively referred to as "open data" under this broad heading. Information on public health ought to go under this heading. Instead of using personal health information, which, if published incorrectly, might breach someone's privacy, public health statistics are often aggregated data that could help in making decisions regarding health-related issues. The benefits and gains that might come from having enough information to make decisions are evidence of the need for readily available public health statistics.

Public health data that is documented and publicly accessible may aid in averting disasters. With a focus on open-source application development and code sharing, a rising number of individuals are advocating for data transparency. In the technology context, the concept of openness has notably gained strength. The need for open access to academic content produced by the scientific community has prompted the establishment of a few initiatives [1]. Government organizations throughout the globe are also developing strategies to improve public use and access to government data [2].

The idea of big data has evolved with the open data movement in the health-care industry. Offering tailored treatment to people, producing early warnings for pandemics, and assisting health system management are just a few of the potential benefits open data platforms bring.

Several sources provide open data on health, including census information, results of surveys on the economy, labor, and education, as well as meteorological information. In addition, on health data files are available on websites such as data.gov, academic journal websites, institutional websites, websites for United Nations agencies, and general-purpose websites [3].

Open data's perspective for improving public health [4].

- Enhances analytical and scientific studies capabilities
- Enhances earlier-than-usual detection and prevention of health and safety risks
- Enhances alternative assessment and monitoring of valid reactions
- Enhances capabilities for increased transparency
- Enhances evaluation capacity and quality measures
- Enhances early surveillance of the environment and wellness risks
- Enhances earlier-than-normal tracking for health allegations.

Enormous resources are needed to shift databases toward becoming open, interoperable, and accessible via common protocols and vocabularies. As opposed to the mere fact that single databases can be used more widely, the capacity to use, exchange, and combine this data with other data is the actual value of open data. A cultural change is also required to move away from the concept of databases as proprietary intellectual property and toward the idea of data as a public good.

With more reliable and accessible data, AI is projected to be a major factor behind analytics, insights, and the decision-making procedure. As a result, oanizations that use AI to change their products and services to increase consumer engagement are likely to enjoy quick returns and sustainable strategic superiority [5].

- The top goals for healthcare firms using AI are to increase process efficiency, improve current goods and services, and reduce costs.

- The expense of the techniques, incorporating AI into the business, and implementation obstacles, including AI hazards and data issues, were cited as the top concerns concerning risks with AI by healthcare organization respondents.

Health systems were overburdened by the current epidemic, which also revealed their shortcomings in terms of providing treatment and controlling expenses. Because of the need and regulatory flexibility, virtual health underwent a historic transition beginning in March 2020. As health systems, health plans, and PBMs develop their new AI investment strategies, analyzing how healthcare companies utilized artificial intelligence (AI) in the aftermath of the pandemic will continue to be helpful. Although the study was carried out before the public health emergency, some of the lessons are still relevant today.

The following are the main characteristics of AI systems:

- AI (artificial intelligence) systems might be able to do a better job than regular computer systems.

- Their fundamental abilities are comparable to human intelligence. Examples of effective patterns include classification, anomaly detection, regression, and prediction.

- The application of these talents to data sets and problems that are far larger.

- More complex than those that people can handle is what distinguishes artificial intelligence (AI) from other technologies as a whole.

The widespread usage of artificial intelligence (AI) and digital gadgets is rife with challenges, including issues with privacy [6, 7], cybersecurity, data integrity, ownership, and sharing. Ethical issues are challenging to overcome in the healthcare industry since AI technology has the potential to compromise patients' autonomy, security, and privacy [8]. The rate of AI development now lags behind the policies and moral guidelines for healthcare services that use AI and its applications.

Public open data advantages

- Enhances accountability and transparency

- Building credibility and reputation

- Growth and innovation

- Increased perception and community involvement

- Knowledge is stored and preserved throughout time

The chapter is structured as depicted below. An assessment of the related topic research is presented in Section 2. POD (public open data) in medicine is discussed in Section 3. Section 4 examines the advantages and disadvantages of using free public data when using AI-based technology. Finally, Section 5 discusses challenges and trends in using AI technologies in healthcare data, followed by concluding remarks with future enhancement.

2. Related work

According to several recent studies, AI is capable of and even superior to, performing critical healthcare tasks such as disease diagnosis. Computers are already better than radiologists at spotting malignant tumors these days, and they can also provide researchers advice on how to create cohorts for costly clinical studies. For many reasons, we do not anticipate that AI will ever totally replace people in the context of large-scale medical processes. In this article, we address the opportunities for AI to improve processes related to providing care and some of the barriers preventing AI from being quickly adopted in the health sector [9].

Current advancements in AI applications for drug design and development suggest that DL approaches in models are becoming more popular. Deep-learning models need substantially more time to train than simpler machine-learning methods do because of the size of the training datasets and the sometimes large number of parameters needed. This might be a serious disadvantage when data is hard to come by. As a result, attempts are being made to reduce the amount of data required for training sets for DL so that it may learn new knowledge using just moderate amounts of current data. This is similar to the human brain learning process and would be helpful in cases where large datasets are scarce, and data collection is time- and resource-consuming, as is often the case with medicinal chemistry and novel drug targets. One-shot learning, lengthy short-term memory, and memory-improving neural networks like the differentiable neural computer are just a few of the novel strategies being investigated [10].

The implementation of various websites, open information repositories like GitHub, or open data portals like Kaggle.com may play a crucial role in AI endeavors. Although making open data alone accessible may help larger companies more since they have access to proprietary datasets to combine with open data sources while smaller companies do not, larger companies may still gain from making open data alone accessible. However, depending on the goals of the data analysis and AI applications, the commercial value would mostly come from fusing this open data with specialized data, such as those originating from the firm itself or obtained through internal operations of the organization or networks. Pushing governments to make the data these systems depend on accessible is one strategy to promote data accessibility and enhance data quality. Many governments rely on algorithms and AI systems to provide public services. There must be a substantial quantity of high-quality data accessible for AI systems to be taught, which is not necessarily the case in all nations [11].

Utilizing this cutting-edge technology makes managing medical care more effective. Because there are benefits to using AI in healthcare, the future is not all upbeat. The appropriate law is not yet fully prepared for this breakthrough, and there are several worries about how AI may operate in terms of doctors' rights and obligations and protect privacy issues. Nevertheless, current laws favor AI, as seen by its application in the global healthcare system. The guidelines for developing technology and health technology goods may be established and used in medical care, as has been shown [12]. This research sought to identify the potential benefits and dangers of AI in the healthcare industry.

3. About public open data in healthcare

Openness may mean various things to different people and organizations, although the phrase "open data" sounds self-explanatory. In addition, there may be notions and principles that only apply to certain industries. For instance, open data must be maintained after it has been distributed, according to the Project Open Data of the United States of America [13]. Although they are often only obligated to store the data for a short while after the project is over, this concept also applies to academic research initiatives.

Large databases of patient data have been amassed since the advent of EMRs, and when taken as a whole, they may be utilized to spot healthcare patterns within various illness areas. Laboratory test results, medical pictures, clinical narratives, and records of diagnoses and actions are all included in the EMR databases. Building prediction models from all this information may assist physicians with diagnosis and other therapeutic decision-making processes. It will be feasible to extract a variety of data, including correlations between past and present medical occurrences and information about connected illness consequences, as AI capabilities develop [14]. The patient may be healthy or not be exhibiting any symptoms while the data is absent, yet it is often missing from hospital visits and data collected between therapies. Such data may be utilized to develop an end-to-end model of both "health" and "disease," to study long-term effects and develop new sickness categories.

Unclean and Disturbed Information Additionally, data may be noisy and inaccurate. For instance, the data's labels or contextual information might be wrong, or the readings themselves could be erroneous. However, the problem of crowdsourcing's noise has not yet been addressed. Because crowdsourcing relies on human judgment to give labels to data, particularly when used for participatory sensing, it may potentially produce noisy data. Although big data is not the only source of dirty and noisy data, the methods for dealing with it may not be well suited to dealing with massive datasets.

Nowadays, data about various aspects of our lives are obtained in a variety of ways, but the techniques and methods used to collect the data may introduce ambiguity. A machine-learning system finds it challenging to conclude such data because of this lack of impartiality. This inherent unpredictability cannot be eliminated even by the most advanced data preprocessing techniques [15]. Once again, this presents particular difficulties for machine learning with big data.

An AI-based technology for complete EMR data analysis is DeepCare. To acquire and maintain events in the memory unit, it makes use of a DDMNN (deep dynamic memory neural network). The system's long-term, short-term memory uses a time-stamped series of events to represent user healthcare routines and sickness trajectories, allowing it to identify long-term dependencies [16]. The DeepCare framework can predict illness development, enable intervention advice, and offer disease prognosis based on EMR databases using the stored data. By examining data from a cohort of diabetic and mental health patients, DeepCare was demonstrated to be able to predict the onset of disease, identify the most effective treatments, and estimate the likelihood of readmission.

The capability of machine learning and artificial intelligence may be unlocked via accessibility to POD (public open data). A study using artificial intelligence that is noteworthy demonstrates how a police department that prioritizes crime prevention utilizes large volumes of open data. The computer created patterns to identify "hotspots," places where certain crimes are expected to occur in the future, based on the instances of crimes (data) and their frequency. These "hotspots" are predetermined geographic locations where the algorithm can make accurate estimations about the kind of crime that could happen and when it is most likely to happen. Various presumptions and trends, such as the fact that cybercriminals often operate in the same location for longer periods, serve as the foundation for these forecasts [17].

Over 80% of the time spent on AI initiatives in India is already spent on data preparation and technical duties. Given the wide variety of demographic, socioeconomic, epidemiological, and climatic conditions, information gathered from these geographies is only marginally useful for informing AI models used in India's many regions. The utilization of data from countries with established open data programs is a choice made by Indian engineers in various circumstances.

4. Challenges and trends of public open data using AI

4.1 Challenges

In the SAR database, which is maintained by the Institute of Cancer Study, information from scientific research is combined with genetic and clinical data from actual patients. AI may be used to help in the identification of new drugs and to scan the scientific literature for relevant research [18]. The development of drugs has been sped up and made more affordable because of the creation of Eve, an artificial intelligence (AI) "robot scientist" [19]. In order to identify potential novel cancer medication targets, artificial intelligence is used in the Institute of Cancer Research's SAR database, which integrates genetic and clinical data from patients with data from academic research [17]. In addition, AI systems employed in healthcare might benefit medical research by assisting in matching appropriate individuals to clinical investigations [20].

A key issue for the future of AI governance will be making sure that AI is developed and used in a way that is open to the public, works for the public good, and promotes and speeds up research. Many of the ethical and social concerns highlighted by the use of AI also apply to other uses of data and healthcare technology. AI usage creates a variety of ethical and social concerns.

4.1.1 Finding data

Governments may be making the data public, but that does not always imply it is simple to discover. Governments often lose reliability when they improve their data skills and create better material. As a result, it is more complicated for users to locate the information and files they need on the website.

4.1.2 Usage of data

Even though governments are making the data public, that does not imply that it is "ready for use." There may be variations in format and other compatibility problems. When we compare open data sets across time, they often contain a varied collection of fields. It makes it difficult for consumers to estimate growth in a crucial field for more than 3 years [21].

4.1.3 Quality of data

Clinics must also deal with difficulties related to adequate collection and usage of data for continuous improvement and discrepancy minimization. In a survey that was conducted by the National Public Health and Hospitals Institute (NPHHI) in 2006, hospitals that collected racial and ethnic data were asked if they used the data to assess and compare the quality of care, the use of medical services, or patient satisfaction across their various patient populations. They were used by fewer than one in five hospitals for any of these purposes [22, 23].

4.1.4 Claims-related and managerial

In addition to clinics and insurance providers, the bulk of the data comes from federal, state, and local government agencies. Documentation of payments made by insured people to the healthcare system or summaries of hospital discharges may be included [24]. The phrase "produced in a clinical environment and supervised by a doctor" refers to a wide variety of data kinds [25].

Repositories and the findings of clinical investigations with public and commercial funding are examples of data. In the course of a clinical study, a lot of data is produced that contains personal information about patients. To acquire and utilize this data, investigators must seek legal authorization.

4.1.5 Electronic health records

(EHRs) may be used by doctors to create customized treatment plans and make diagnoses. To create longitudinal profiles of people and populations, this data may also be integrated with socioeconomic determinants of healthiness. EHR data focuses on specific individuals and may include details on regular visits, treatments, and diagnostic interventions.

Genomic data may include a wide range of features, from whole DNA sequences to specific DNA variations. The development of improved, effective treatments, more effective diagnostic tests, evidence-based methods for proving a clinical success, and better tools for patients and providers to make decisions are already made possible by genome-based research [26]. The whole genomic sequence of an individual may now be analyzed and stored as data [27].

4.1.6 The term "patient-generated data"

It refers to health-related information created and recorded by or from patients outside of a therapeutic setting. Due to the development of wearable health technology and mobile health apps, this sort of data is becoming more and more common. Genomic information is regarded as being very sensitive and should only be exchanged and utilized under strictly regulated circumstances [25].

4.1.7 Data from wearable technology

Such as smartwatches, voice assistants, and mobile software apps, is included in IoT data. These data have the potential to provide crucial details on several vital health markers, including heartbeats and sleep patterns. These innovations are components of the expanding network of machines and gadgets linked to the internet known as the "internet of things," or IoT.

4.1.8 Data from social media

Covers communications on websites like Facebook and Twitter. It may shed light on perceptions about wellness as well as the connection between a person's health and their daily lives, according to investigators. Social media data is gathered by "terms of service" contracts, much like IoT data [28].

Health disparities in the demographic health survey relate to "cases in the settings in which people are born, live, learn, and work that affects a wide range of health, functional, and quality-of-life consequences and dangers." Access to food and housing alternatives, as well as possibilities for education and employment, are a few examples of these social determinants [29]. Data on social determinants of health may be obtained from a variety of sources, both within and outside of government, and utilized to improve healthcare quality.

According to the US Department of Health and Human Services, data is the "continuous, systematic collection, analysis, and interpretation of health-related data vital to the planning, implementation, and assessment of public health practices" (HHS) [30]."

When humans use artificial intelligence to uncover similarities in genetic data versus very private patient history data, they face comparable difficulties with data presentation and perspective. Fractal representation and even statistical patterns are incredibly challenging concepts for people to grasp. The computer will be able to characterize anything we observe that has a statistical quality, but we can too [31].

The use of AI to enhance healthcare now faces several challenges for investigators and clinicians. Access control restrictions, data governance issues, and ethical data usage are a few of them.

High-quality, accurate, and clean data are essential to artificial intelligence. Large amounts of health data, including those from wearable technology and sensors as well as electronic health records (EHRs), are now being mined by researchers [32]. Future uses of disruptive AI will be made possible by the healthcare system's increased connectivity and interoperability of data.

In the area of digital forensics, there is a severe lack of professionals. A novel strategy is required because of the rising demand. The solution to this demand gap may be found in artificial intelligence. It can decipher data presented as either an image or a video.

The data produced by clinical procedures, including medical examinations, diagnostics, therapy assessments, and other similar clinical activities, must first be "trained" into AI systems before they can be utilized in healthcare systems. This enables them to recognize subject groupings that are comparable to one another and establish relationships between subject characteristics and desired results. These clinical data often take the form of demographic information, medical records, electronic recordings from medical equipment, physical examinations, clinical laboratory findings, and photos, but they are not just these [33].

A further issue is the dissemination of information. Clinical trial data must be frequently used to train AI systems if they are to work successfully. Maintaining the data source, however, becomes a critical challenge for the system's continuing development and improvement whenever an AI system is placed into service after its first training on historical data. Incentives for exchanging system data are currently missing from the healthcare perspective. However, the US healthcare industry is undergoing a transition that will encourage data sharing. A new payment system for health services [34] is where the reform process begins. Numerous payors—mostly insurance companies—have switched from compensating doctors by changing the amount of their patient care to rewarding them for the quality of their care.

In the past, personal health records have often lacked patient-related functions and have been more physician-focused But a patient-centered personal health record is a must if we want to encourage people to take care of themselves and improve patient outcomes. While giving professionals more time to focus on more pressing and important responsibilities, the aim is to give patients more independence to control their diseases.

A healthcare AI algorithm's development requires a specialized dataset, which presents a challenge. Because of this, the resulting data model may not precisely represent local patient data. In addition, the clinical and ethical concerns in various medical specialties, like radiology or pediatrics, vary, so it is important to analyze the dangers of AI in the context of each relevant area [35].

Designing new, safe computer system ecosystems and rethinking how we perceive privacy and control is also necessary for data security in massively dispersed infrastructures. Large datasets with high noise are a significant issue that calls for novel analytics techniques. The amount of resilience required by the techniques now in use leads to errors and the production of false positive signals [36].

A challenge arises when research findings from huge databases are utilized to select user-served persons and disease regions for treatments before the availability of scientific proof. Providing sample data that is accessible to the public might give hackers access to vulnerable AI models. Attacks may exploit data poisoning when open datasets are generated by the public or subject to public updates. In one research study, the danger posed by adversarial assaults that slightly alter pictures for medical imaging software was investigated. Although these changes were not evident to the human eye, deep learning algorithms may nonetheless misclassify images up to 100% of the time [37]. This kind of attack may have severe ramifications since several organizations, including government organizations, provide public databases of medical images to assist in diagnosis and treatment [38].

The majority of individuals believe that AI technology will enhance and assist human labor rather than replace physicians and other care providers. AI is ready to help healthcare personnel with a variety of tasks, including administrative operations, clinical notes, and patient engagement. Additionally, it may provide specific assistance in patient monitoring, picture analysis, and the automation of medical devices.

The challenge of monitoring in compartmentalized EHRs may be resolved by using AI techniques, which will reroute such reports to analysis and predictive modeling. Programs for preventive healthcare may also use this technology. For example, it may integrate data from various data sources, such as electronic health records (EHRs), with a person's omic (genome, proteome, metabolome, and microbiome) data to predict the likelihood of getting a disease [39].

Utilizing AI to assess clinical information, scholarly publications, and ethical standards may help in making decisions about the treatment to provide patients [40].

4.2 Trends

Each year, thousands of hospital patients experience avoidable suffering and death as a result of medical mistakes. These mistakes are often caused by doctors handling a heavy caseload with insufficient medical histories. Faster than most medical practitioners, AI can forecast and diagnose illnesses. Incorporating AI into EHR software has been a gradual process for companies [41].

4.2.1 AI in administration

For the handling of administrative data, many healthcare companies are turning to AI. AI can speed up and reduce error rates in a variety of administrative tasks, including insurance processing, clinical notes, management of revenue cycles, healthcare document management, and some other administrative functions. Utilizing AI solutions to identify and correct code defects and false claims might result in significant attention, revenue, and manpower savings for these businesses.

4.2.2 Drug discovery and AI

In terms of drug research and clinical development, AI represents a paradigm shift. The discovery process may be sped up significantly by utilizing AI's effectiveness, precision, and rate of data processing. Depending on the medicine class, just the clinical studies themselves might cost millions of dollars. Even after the clinical testing stage, only 10% of medications reach the market.

4.2.3 AI for healthcare and robotic surgery

Robotic surgical suites driven by AI provide clinicians with a high level of accuracy, flexibility, and control than they would otherwise have. Robot-assisted operations, therefore, result in fewer surgical problems, reduced postoperative discomfort, and quicker recovery durations. Even future doctors will be trained with surgical robots.

4.2.4 Biassed data

Several factors, such as social prejudice (inaccessibility to healthcare) and small populations (for instance, minority communities), might lead to the existence of data that is not reflective [42]. For the AI model's training, a sizable quantity of information regarding health data or other topics is needed.

4.2.5 Organ care improvement

One way to increase organ care is to take care of the organ while it is outside the body. In addition, machine learning may allow for a more precise evaluation of a preserved organ's transplant ability. If this could be found more quickly, lives could be saved more quickly.

4.2.6 Bioprinting

Other alternatives should be investigated besides keeping organs alive outside the body; although they seem like science fiction, 3D-printed organs are a very real, still emerging technology that has already entered clinical testing. Clinical trials for 3D bioprinting of bones, skin, corneas, ears, and other organs are underway.

The method of developing a digital organ model that can be printed out is called bioprinting. Bioink, which is made of live cells, is the ink used in printing. By analyzing organ and patient traits using AI, it is possible to better design organs to be compatible with their new hosts. In every way, medical technology will develop. Threats always evolve and must be dealt with via prevention rather than response, despite industry-wide security advancements.

4.2.7 Data privacy protection

The main foundation upon which DL and ML models are formed is the availability of data and resources to train these models. Given that this data is generated by millions of individuals throughout the globe, there is a chance that it may be abused. In other firms, creative efforts to overcome these challenges have already started. On smart devices, the data is used to train the model, and only the trained model is delivered, with no training data being sent back to the servers [43].

4.2.8 Limited data

AI heavily relies on data, and annotated data is used to teach computers to understand and predict. Numerous companies are putting their attention toward developing artificial intelligence (AI) solutions that can provide reliable results in the absence of data. The whole framework might become unreliable due to inaccurate data.

4.2.9 Regulatory concerns that arise with AI

Regarding the use of AI in medicine, there are presently no internationally uniform legislation or regulations [44]. AI crime, a brand-new, harmful crime, might happen if criminals employ AI [45]. Legal professionals cannot create such laws on their own. Participants interested in the implementation of AI-based treatment modalities must be addressed [46].

4.3 Medical industry developments including AI

- Healthcare AI,
- Telemedicine,
- Remote treatment development,
- Augmented awareness in healthcare environments,
- Bioprinting and innovation for organ care,
- Smart and IoT in Healthcare,
- Healthcare data protection.

Another challenge is the limitation of intent [47]. Data may be used for purposes other than those for which they were intended if copyright rules are not drafted properly [48]. Regarding the repercussions of decisions taken by automated decision-making systems, authorities must be clear about who is responsible for implementing AI [49].

In terms of data availability and value, open source presents unique issues. Transparent and excellent datasets are not always a result of open-source technologies. To understand the idea of accessibility, the whole open-source ecosystem must be taken into account. There is still a lack of transparency on how AI quality can be evaluated when using open-source AI software on "closed data." Consequently, the potential advantages of open source are hampered since not all components connected to AI are open source. Adopting standard protocols is essential to achieving true transparency.

Due to its capacity to promote openness and accountability in government, open data may help address some of these issues. If these programmers are volunteers, they are allowed to leave the project anytime they choose, which might impede its growth since the work's quality and productivity rely on continued developer engagement. The decentralized AI strategic plan is one potential solution to address issues related to data confidentiality and risks while keeping open sources.

Technology for patient care support may increase physicians' workload and promote the mobility and health of patients. For instance, remote medical assistants may suggest different exercise routines or remind people to take their prescribed drugs at certain times. In addition, patients will have remote access to the gadget and be able to view their biometric data while still feeling as if communicating with a kind and sympathetic system.

To aid patients with lifting and transporting heavy objects, the RIBA assistive robot—named for its human-like arms—was created. The use of tactile sensors or tactile guidance to explain instructions to RIBA is also an option. Robotic patient transfer from a bed to a wheelchair and vice versa is possible [50].

The patient can provide as much information as they want in text, photographs, audio, and video. This enables the doctor to examine and evaluate the data before speaking with the patient. This is quite inspiring and inventive, considering how many individuals do not have the time or money to visit a doctor.

4.4 Advantages of AI in healthcare data

4.4.1 Information access

Providing quality data in real-time is one of artificial intelligence's stronger points in the health sector. Faster diagnosis based on the findings is made possible, which significantly positively impacts patients' chances of recovering or following their treatment plan. Medical practitioners may also get real-time information on the status, crises, and changes that the patient may have experienced via smartphone notifications.

4.4.2 Challenge optimization

AI has helped the healthcare sector simplify a lot of work, including scheduling visits and transferring patient information and medical histories. Radiotherapy uses advanced analytics that can even intuitively pick out key signs. This makes it faster to look at diseases in depth.

4.4.3 Expenditure and insight

The costs of clinics might be significantly lowered when AI replaces laborious human work with sophisticated algorithms. Certain AI can also help with evaluations to provide an analysis of what the clinic requires.

4.4.4 Investigative competence

AI may incorporate different information sources based on the study, which can be very useful for assessing illnesses, other than supplying real-time data. To aid in the essential processes and alternatives per developmental stage, software to treat certain major illnesses has been developed.

5. Discussion

The potential for releasing healthcare data and sharing large datasets is immense, but significant obstacles and hurdles exist. Data visibility issues and a lack of connections between individuals and organizations working to tackle comparable issues are significant hindrances to global development. Nevertheless, collaboration would be accelerated, and some of the largest problems in global health would be resolved via a sustained open data revolution. Other difficulties include combining the need to promote innovation with ensuring that AI development and usage are open, responsible, and consistent with the public interest. Numerous people have stressed the need for academics, medical practitioners, and policymakers to have the abilities and expertise to assess and use AI to its fullest potential [51].

Effective health treatment choices and outcomes are only one benefit of using medical technology. Additional benefits might include reduced admissions, cheaper costs, and much more effective provision of resources. It may also assist local health centers and entice patients to live and operate therein. A further equal universal medical system may result in the end [52]. Boosting quick acceptance, sustainable implementation in the health system, a lack of respect for user perspectives, and the impossibility of automation being utilized to its complete capacity without the adoption of AI in the public health sector are the challenges. A foundation for future studies on many facets of digital innovation in the health sector is provided by the usage of AI in health service [53].

6. Conclusion

Several adjustments would need to be made to use open data in the healthcare system fully. The most basic one would be a move to data-driven techniques in health and care, where medical choices regarding treatment are based on data from thousands of individuals. The full potential of open data in healthcare is also hindered by organizational and technological issues, such as the difficulty of many healthcare data systems to provide consistent data. Finally, owing to the repercussions of improper handling of medical data, privacy and confidentiality are issues. Nevertheless, understanding, controlling, and reducing the consequences of health issues on people, society, and the economy depends on data.

The issues and challenges include stimulating early acceptance, long-term deployment in the healthcare system and disregard for the viewpoint of the user. Despite not being exploited to its full potential, technology is essential for the advancement of artificial intelligence in the medical sector. Developing new approaches to a healthy alignment involves identifying innovative ways to meet corporate needs and effectively involve the public. Large-scale integration and accessible data access for scientific purposes while protecting privacy rights.

7. Future enhancement

In the future, we will focus on analyzing data infrastructure, effective information exchange and integration, and efficient data release in open and machine-readable forms. The fact that there is a different balance between privacy and openness adds to the difficulty of mitigating the danger of re-identification of quality data. As public-private data collaborations become more significant, it is crucial to make sure that these initiatives promote equitable growth. Although the data protection and privacy regulations only make up a small portion of a country's overall data governance system, they often get the most attention since they deal with politically touchy subjects. In this chapter, we discussed the various trends and challenges of open data, as well as how AI technologies use these data. Using public open data, we will create some AI real-time applications.

Author details

Vijayalakshmi Kakulapati
Sreenidhi Institute of Science and Technology, Hyderabad, Telangana

*Address all correspondence to: vldms@yahoo.com

References

[1] Stebbins M. Expanding Public Access to the Results of Federally Funded Research. Available from: https://www.whitehouse.gov/blog/2013/02/22/expanding-public-access-results-federally-funded-research

[2] Sheehan J. Increasing Access to the Results of Federally Funded Science. Available from: https://obamawhitehouse.archives.gov/blog/2016/02/22/increasing-access-results-federally-funded-science

[3] Available from: https://globalhealthdata.org/opening-health-data-to-the-public

[4] Huston P, Edge VL, Bernier E. Reaping the benefits of Open Data in public health. Canada Communicable Disease Report. 2019;**45**(11):252-256. DOI: 10.14745/ccdr.v45i10a01

[5] Available from: https://www2.deloitte.com/us/en/insights/industry/health-care/artificial-intelligence-in-health-care.html

[6] Safavi K, Kalis B. How AI can change the future of health care. Harvard Business Review. 2019. Available from: https://hbr.org/webinar/2019/02/how-ai-can-change-the-future-of-health-care

[7] Yoon S, Lee D. Artificial intelligence and robots in healthcare: What are the success factors for technology-based service encounters? International Journal of Healthcare Management. 2019;**12**:218-225

[8] Rigby M. Ethical dimensions of using artificial intelligence in healthcare. AMA Journal of Ethics. 2019;**21**:E121-E124

[9] Davenport T, Kalakota R. The potential for artificial intelligence in healthcare. Future Healthcare Journal. 2019;**6**(2):94-98. DOI: 10.7861/futurehosp.6-2-94

[10] Lavecchia A. Deep learning in drug discovery: Opportunities, challenges, and future prospects. Drug Discovery Today. 2019;**24**(10):2017-2032

[11] Theben A, Gunderson L, López Forés L, Misuraca G, Lupiáñez Villanueva F. Challenges, and limits of an open source approach to Artificial Intelligence, study for the Special Committee on Artificial Intelligence in a Digital Age (AIDA), Policy Department for Economic, Scientific, and Quality of Life Policies. Luxembourg: European Parliament; 2021

[12] Faridah L, Rinawan FR, Fauziah N, Mayasari W, Dwiartama A, Watanabe K. Evaluation of Health Information System (HIS) in The Surveillance of Dengue in Indonesia: Lessons from Case in Bandung, West Java. Int J Environ Res Public Health. 10 Mar 2020;**17**(5):1795. DOI: 10.3390/ijerph17051795

[13] Project Open Data Principles. Available from: https://project-open-data.cio.gov/principles/

[14] Pham T, Tran T, Phung D, Venkatesh S. Predicting healthcare trajectories from medical records: A deep learning approach. Journal of Biomedical Informatics. 2017;**69**:218-229

[15] Hausenblas M, Jacques Nadeau J. Apache Drill Adhoc interactive analysis at scale. Jun 2013. DOI: 10.1089/big2013.0011

[16] Pham T, Tran T, Phung D, Venkatesh S. DeepCare: A deep dynamic memory model for predictive medicine.

In: Lecture Notes in Computer Science (including subseries Lecture Notes in Artificial Intelligence and Lecture Notes in Bioinformatics). Springer; 2016

[17] Available from: https://data.europa.eu/en/publications/datastories/ai-and-open-data-crucial-combination

[18] O'Mara-Eves A et al. Using text mining for study identification in systematic reviews: A systematic review of current approaches. Systematic Reviews. 2015;**4**:5

[19] The Conversation (11 November 2013) Artificial Intelligence uses the Biggest Disease Database to Fight Cancer

[20] Williams K et al. Cheaper faster drug development validated by the repositioning of drugs against neglected tropical diseases. Journal of the Royal Society Interface. 2015;**12**:20141289

[21] Alder Hey Children's NHS Foundation Trust. 2016. Alder Hey Children's Hospital is set to Become UK's First 'cognitive' Hospital

[22] Regenstein M, Sickler D. Race, Ethnicity, and Language of Patients: Hospital Practices Regarding the Collection of Information to Address Disparities in Health Care. Princeton, NJ: Robert Wood Johnson Foundation; 2006

[23] Hasnain-Wynia R, Pierce D, Pittman MA. Who, When, and How: The Current State of Race, Ethnicity, and Primary Language Data Collection in Hospitals. New York: The Commonwealth Fund; 2004

[24] University of Washington Health Sciences Library, "Data Resources in the Health Sciences". Available from: http://guides.lib.uw.edu/hsl/data/findclin

[25] Office of the National Coordinator for Health Information Technology, Conceptualizing a Data Infrastructure for the Capture, Use, and Sharing of Patient-Generated Health Data in Care Delivery and Research through 2024, 2018. Available from: https://www.healthit.gov/sites/default/files/onc_pghd_final_white_paper.pdf

[26] National Institutes of Health National Human Genome Research Institute, "A Brief Guide to Genomics". Available from: https://www.genome.gov/about-genomics/fact-sheets/A-Brief-Guide-to-Genomics

[27] PHG Foundation at the University of Cambridge, Identification and Genomic Data, 2017. Available from: http://www.phgfoundation.org/documents/PHGF-Identification-and-genomic-data.pdf

[28] Padrez K et al. Linking social media and medical record data: A study of adults presenting to an academic, urban emergency department. BMJ Quality & Safety. 2016. DOI: 10.1136/bmjqs-2015-004489

[29] Office of Disease Prevention, and Health Promotion, "Social Determinants of Health". Available from: https://www.healthypeople.gov/2020/topics-objectives/topic/social-determinants-of-health

[30] World Health Organization, "Public Health Surveillance". Available from: https://www.who.int/topics/public_health_surveillance/en/

[31] Available from: https://emerj.com/ai-sector-overviews/healthcare-ai-use-cases-and-trends-an-executive-brief

[32] Ismail N. The success of artificial intelligence depends on data. Information Age. 2018. Available from: https://www.information-age.com/success-artificial-intelligence-data-123471607/

[33] Available from: https://www.federalregister.gov/documents/2013/09/18/2013-22645/guidance-for-industry-on-electronic-source-data-in-clinical-investigations-availability

[34] Kayyali B, Knott D, Kuiken SV. The Big-data Revolution in US Health Care: Accelerating Value and Innovation. 2013. Available from: http://www.mckinsey.com/industries/healthcare-systems-and-services/our-insights/thebig-data-revolution-in-us-health-care

[35] Available from: https://www.computerweekly.com/news/252524829/The-challenges-of-verifying-AI-for-healthcare

[36] Havens JC. Artificial Intelligence is Doomed if We Don't Control our Data. 2014. Available from: http://mashable.com/2014/09/16/artificial-intelligence-failure/

[37] Choi CQ. Medical imaging AI software is vulnerable to covert attacks. IEEE Spectrum. 2018

[38] Kent J. NIH makes the largest set of medical imaging data available to public. HealthITAnalytics. 2018

[39] R. Eubanks, 2017, Artificial Intelligence and the Healthcare Ecosystem – Part One. Available from: https://www.capgemini.com/2017/10/artificial-intelligence-and-thehealthcare-ecosystem-part-one/. [Accessed: January 5, 2018]

[40] Dilsizian SE, Siegel EL. Artificial intelligence in medicine and cardiac imaging. Current Cardiology Reports. 2013;**16**:441

[41] Future Advocacy. Ethical, Social, and Political Challenges of Artificial Intelligence in Health. 2018. Available from: https://taazaa.com/healthcare-ai-trends/

[42] Akmal A, Greatbanks R, Foote J. Lean thinking in healthcare – findings from a systematic literature network and bibliometric analysis. Health Policy (New York). 2020;**124**:615-627. DOI: 10.1016/j.healthpol.2020.04.008

[43] Available from: https://www.upgrad.com/blog/top-challenges-in-artificial-intelligence/

[44] Mitchell C, Ploem C. Legal challenges for the implementation of advanced clinical digital decision support systems in Europe. Journal of Clinical and Translational Research. 2018;**3**:424-430

[45] King TC, Aggarwal N, Taddeo M, et al. Artificial intelligence crime: An interdisciplinary analysis of foreseeable threats and solutions. Science and Engineering Ethics. 2020;**26**:89-120

[46] Cath C. Governing artificial intelligence: Ethical, legal and technical opportunities and challenges. Philosophical Transactions of the Royal Society A. 2018;**376**:20180080

[47] Available from: https://mobidev.biz/blog/technology-trends-healthcare-digital-transformation

[48] Finland AI. 3 Eleven Key Actions Ushering Finland into the Age of Artificial Intelligence. AI Finland; 2019

[49] Lee J. Government policy toward open-source software: The puzzles of neutrality and competition. Knowledge, Technology, & Policy. 2006;**18**:113-141

[50] Sweden AI. Decentralized AI. AI Sweden; 2021

[51] Joseph A, Christian B, Abiodun AA, Oyawale F. A Review on Humanoid Robotics in Healthcare. MATEC Web of Conferences; 2018. Available from: https://www.matec-conferences.org/

[52] Available from: https://www.openaccessgovernment.org/what-are-the-pros-and-cons-of-implementing-ai-in-healthcare/140058

[53] Powles J, Hodson H. Google DeepMind and healthcare in an age of algorithms. Health Technology. 2017;7:351-367

Chapter 2

Pandemic Open Data: Blessing or Curse?

Claus Rinner

Abstract

The SARS-CoV-2 pandemic spawned an abundance of open data originally collected by local public health agencies, then aggregated, enriched, and curated by higher-level jurisdictions as well as private corporations such as the news media. The COVID-19 datasets often contain geospatial references making them amenable to being presented cartographically as part of map-centered dashboards. Pandemic open data have been a blessing in that they enabled independent scientists and citizen researchers to verify official proclamations and published narratives related to COVID. In this chapter, however, we demonstrate that these data also are cursed with serious issues around variable definitions, data classification, and sampling methods. We illustrate how these issues interfere with unbiased public health insights and instead support narratives such as the "pandemic of the unvaccinated." Nevertheless, open data can serve as a tool to counter dominant narratives and state-sanctioned misinformation. To advance this purpose, we need to demand disaggregated data with transparent metadata and multiple classification schemes.

Keywords: COVID-19, citizen research, geospatial data, misinformation, narratives, post-pandemic recovery

1. Introduction

Along with the SARS-CoV-2 pandemic, we experienced the emergence of a tremendous amount and variety of open data collected, shared, and used globally. However, upon closer inspection, many of the publicly available datasets are marred by data quality and semantic issues. This unique and still evolving situation motivated the critical analysis of COVID-19 open data presented in this chapter. Awareness of these issues is of critical importance for analysts and decision-makers to avoid repeating past mistakes in future public emergencies.

Open data and open content are defined by the Open Knowledge Foundation as those which "can be freely used, modified, and shared by anyone for any purpose" [1]. This includes data originating from government agencies, research institutions, and other organizations, including private-sector corporations as well as non-profits. The idea behind open data is to promote transparency, accountability, participation, and innovation in society by making it easier for individuals and groups to access and utilize data for various purposes. Open data are also supposed to facilitate communication and collaboration between citizens and governments [2].

Many organizations and initiatives are dedicated to promoting and implementing open data, such as the Open Data Institute, the World Bank, and the European Commission. Open data initiatives are becoming more common around the world. They are intended to support a wide range of applications such as journalism, civic technology, higher education, and scholarly research. These initiatives have the potential to create economic value by promoting innovation, efficiency, and competition. Using freely available open data, individuals and organizations can develop new products and services, improve existing ones, and make data-driven decisions. Yet, producing and consuming open data are not always trivial tasks: "Infomediaries, both human and nonhuman, negotiate the gap between open data providers and end-users, and can take the form of service providers, portals, and platforms" [3].

In a municipal context, it was found that some 80% of open datasets include geographic information [4]. These geospatial open data can be used for cartographic mapping and location-based analyses. Geospatial open data add value to economic development, urban and infrastructure planning, and a number of other areas of government activities with citizen involvement [5]. Gig economy companies like Uber and Airbnb have been able to leverage geospatial open data on transportation and housing, respectively, in order to create entirely new markets. Organizations and initiatives dedicated to promoting and implementing geospatial open data include OpenStreetMap Foundation, Open Source Geospatial Foundation, UN Open GIS Initiative, and US Geological Survey.

Medicine and public health are considered among the most promising applications of open data because of the potential of health information to improve patient outcomes and healthcare costs. Examples of open health data include clinical trial results, epidemiological data, healthcare system monitoring, and more. During the COVID-19 pandemic, open data in all of these areas of public health became almost ubiquitous in the media and social media. Most of the datasets include information about point locations or spatial units and can therefore be mapped and spatially analyzed. Numerous map-centered dashboards are a testament to this phenomenon [6, 7].

Nevertheless, open data in health also come with concerns related to privacy, security, and the ethical use of data. The widespread use of COVID data by professional and lay analysts has exposed serious issues with variable definitions, sampling, and categorization. In the following section, we review the past three years with respect to the most critical COVID data issues. Next, we briefly exemplify open data repositories and some of the best-known online applications using these data. In Section 4, we illustrate the impacts of open data issues on four recurring narratives of the pandemic. Lastly, we conclude the chapter with recommendations that could help ensure that open data are indeed a blessing in future crisis situations.

2. The pandemic data circus

The term "data circus" could be used to refer to a situation where large amounts of data are being collected, analyzed, and disseminated in an uncoordinated and chaotic manner, with little regard for quality or accuracy. During the COVID-19 pandemic, data have been employed to track the spread of the virus and understand its impact on communities, as well as to inform public health decisions and policies.

Open data of interest for this analysis can be grouped by interventions, outcomes, and contributing factors, and include the following:

- Confirmed cases and deaths: These data include information on the number of confirmed cases and deaths from COVID-19, and are used to track the spread of the virus and the impact of public health interventions.

- Testing and test positivity: These data include information on the number of tests being conducted and the percentage of tests that are positive, and are used to understand the spread of the virus and the effectiveness of testing strategies.

- Health system capacity and usage: These data include information on the number of people who have been hospitalized and the number of people in ICU beds, and are used to understand the impact of the virus on the health system and the readiness of the health system to respond to the pandemic.

- Vaccination rate and distribution: These data include information on the number of people who have been vaccinated and on the availability of vaccines, and are used to understand the progress of vaccination campaigns and the potential impact of vaccines on controlling the spread of the virus.

- Safety and efficacy of non-pharmaceutical interventions and vaccines: These data include the timelines of interventions such as mask mandates, the SARS-CoV-2 epidemic curves, as well as the number and ratios of adverse event reports, and are used to understand the benefits and risks of these measures.

- Socioeconomic and public health statistics: These data often originate from the Census of the respective populations and from other general-purpose sources such as geodemographics, tax databases, or public health monitoring.

A number of researchers have criticized the accuracy of COVID-related data or the way in which these data were interpreted for public health decision making during the pandemic response. John Ioannidis is a professor of medicine and statistics at Stanford University, who is known for his research on the reliability of scientific studies. He has been critical of how COVID-19 data were reported and analyzed from the beginning of the pandemic, which he called "a fiasco in the making" [8]. He highlighted the age distribution and comorbidities of the fatalities in the early COVID-19 hotspots around the globe such as Northern Italy [9] and New York [10]. He also took it upon himself to create more accurate and reliable epidemic data in a case study for Santa Clara County, California [11]. His low estimates of the global infection-fatality rate for COVID, which were eventually published in the Bulletin of the World Health Organization (WHO) [12], were highly controversial. Ioannidis was also one of the first to explicitly warn of "the harms of exaggerated information and non-evidence-based measures" [13].

Prof. Carl Heneghan of the University of Oxford together with his colleagues at the Centre for Evidence-Based Medicine also published a series of critiques of public health practices that affected the COVID data, including the value of a positive PCR test [14] and definition of a COVID death [15]. Similar concerns were raised by another British research group around Prof. Norman Fenton of Queen Mary University. Fenton contributed to research on face mask efficacy [16] and the reliability of vaccine adverse event reports [17]. On their blog, Fenton and his colleague Prof. Martin Neil further explain the "Flawed Covid definitions, data and modelling" [18] that are pertinent to this chapter.

In Germany, a group around psychology professor Christof Kuhbandner at the University of Regensburg pointedly analyzed and commented on the impact of poor-quality data on the validity of policy decisions on pandemic response measures [19]. Similarly and more specifically, a team led by another psychologist, Prof. Oliver Hirsch of FOM University of Applied Sciences Siegen, found that mass testing yields unreliable data for COVID incidence calculations, which nevertheless were used for policy-making [20]. Even the mRNA vaccine trial data were not safe from criticism and review. For example, the British Medical Journal reported a whistle-blower's concern about improper monitoring and follow-up on vaccine adverse events in the Pfizer trial [21], which could lead to embellished adverse event data.

Some of these researchers have been censored when their viewpoints diverged too far from the prevailing narrative; the tactics used by news media and big tech companies to suppress such dissenting opinions are outlined by Shir-Raz et al. [22]. Questions about the integrity of the scientific and scholarly publication process have emerged as well (e.g., [23]). On the upside, the widespread availability of COVID-19 data, and the ability to force the disclosure of additional datasets, allowed scholars, professional analysts, and private citizen researchers to verify/falsify claims made by public health agencies, pharma corporations, and established experts.

3. Select open data repositories and apps for COVID-19

Many COVID data trackers and repositories in the past three years were created to track the spread of the virus, the impact of the pandemic, and the effectiveness of interventions. Whether based on government and international organizations, research institutions, or the private sector (e.g., news media corporations), foundational data have been provided freely during the course of the pandemic with few strings attached. Most of the following repositories are wrapped in a graphical user interface that allows the user to explore select data before downloading them.

National and international government organizations offering COVID-related open data include the World Health Organization (WHO), US Centers for Disease Control and Prevention (CDC), and European Center for Disease Prevention and Control (ECDC), to name a few. The WHO's global "Coronavirus (COVID-19) Dashboard" [24] presents key statistics on the COVID-19 pandemic, including cases and deaths as cumulative totals or rates per 100,000 population as well as recent cases and deaths, along with several vaccination-related variables, including doses administered, persons vaccinated in percent of total population, and vaccine brand(s) used in different countries. Unfortunately, the default view of the dashboard uses choropleth symbology, a cartographic technique that misrepresents raw-count data such as the cumulative COVID-19 cases shown (**Figure 1**). On a positive note, the WHO datasets can be accessed as comma-separated value (CSV) files in three clicks from the dashboard, with metadata on the field name, type, and description provided prior to download.

The ECDC's "Latest COVID-19 Data" website [25] includes weekly (until October 2022: daily) and cumulative case and death counts, corresponding 14-day rates, as well as information on testing, SARS-CoV-2 variants in circulation, hospital and ICU admissions and occupancy, and vaccination. However, these data for European Union countries are copyrighted (not open). In contrast, the United States CDC's "COVID Data Tracker" [26] presents a large variety of data, charts, and maps at the state or county levels. The available data include variables such as new weekly COVID-19 cases and deaths as well

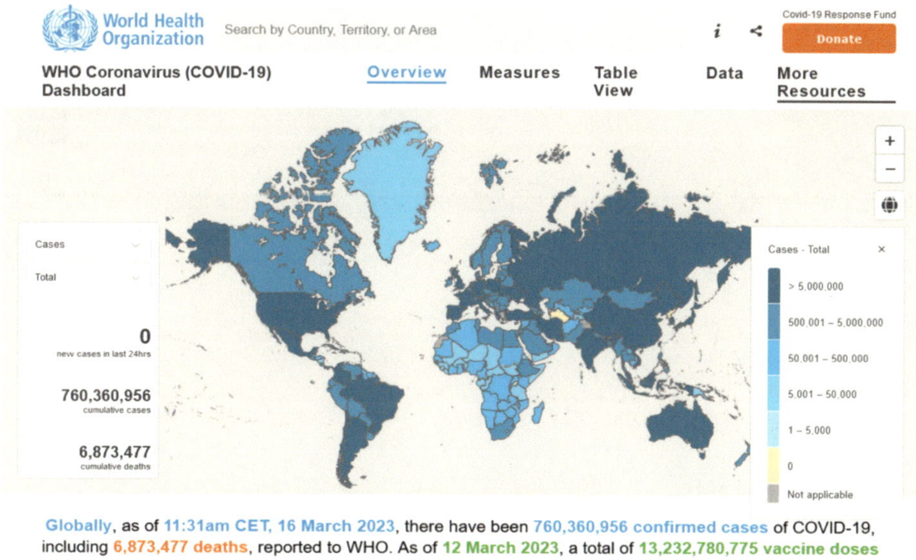

Figure 1.
Screenshot of WHO COVID-19 dashboard as of March 16, 2023. Note that the default view (choropleth map for raw-count data) conflicts with basic cartographic standards and the map is not properly projected. Source: https://covid19.who.int/, used with permission.

as hospital utilization; metrics such as community levels and community transmission; and related information such as the CDC's Social Vulnerability Index. The corresponding data tables are extensively documented and can be downloaded in multiple data formats under the US government's public domain license.

Examples of additional COVID-related government open data sites include other national sites such as that of the Public Health Agency of Canada [27], which includes wastewater surveillance information with downloadable "viral load" data; the UK's Office for National Statistics [28] with an extensive collection of COVID-related content that is "available under the Open Government Licence v3.0, except where otherwise stated"; the COVID-19 landing page of the German infectious disease agency RKI [29] with categorized content on current epidemiology, vaccination, diagnostic testing, healthcare and therapy, long COVID, white papers and pandemic plans, research, and open data. Subnational sites such as that of Public Health Ontario often include more specialized data, for example, regional trends by public health unit, cases associated with long-term care home outbreaks, or neighborhood diversity and material deprivation [30]. The US Vaccine Adverse Event Reporting System [31] is another example of a separately collected, specialized dataset.

Numerous academic and private-sector websites serve as aggregators and curators of pandemic open data. Some of them have become better known than their original government data sources, including Worldometers' COVID Live—Coronavirus Statistics [32], Our World in Data (OWID) [33], the COVID-19 Data Repository at the Center for Systems Science and Engineering (CSSE) at Johns Hopkins University [34, 35], The Atlantic's COVID Tracking Project (discontinued March 7, 2021) [36], and Esri Canada's COVID-19 Open Data hub [37].

Using the example of OWID, **Figure 2** illustrates the central role of maps along with associated charts, data tables, and download options for the open datasets. The OWID site also provides an example of processed information, as it includes

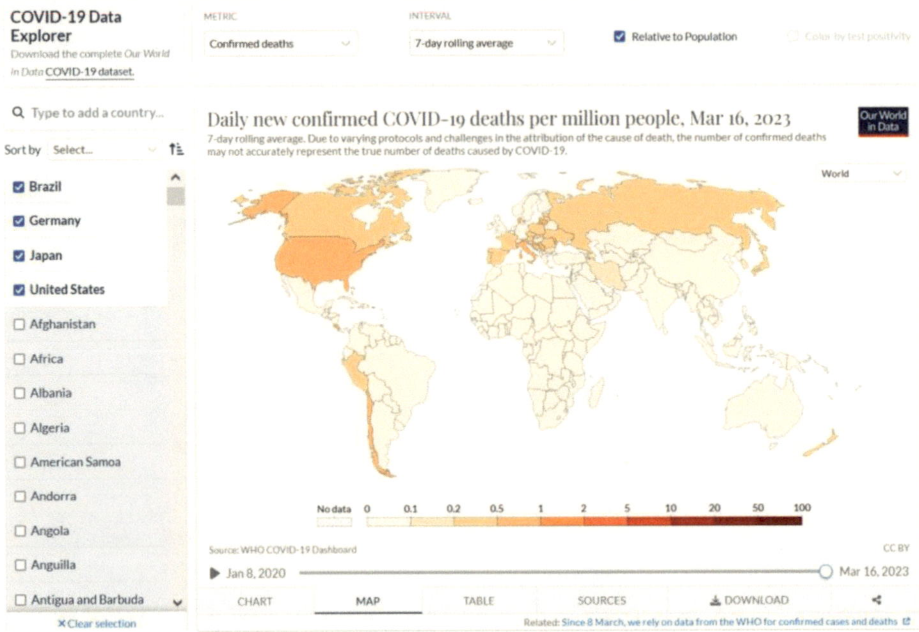

Figure 2.
Screenshot of Our World In Data, Coronavirus Pandemic (COVID-19) website as of March 21, 2023. Source: https://ourworldindata.org/coronavirus, used under Creative Commons BY license.

the Stringency Index from the Oxford Coronavirus Government Response Tracker (OxCGRT) [38]. That project uses open content on pandemic restrictions to model the stringency of national response measures in a single metric. Another example of an application based on processed data is the website "How Bad is My Batch," [39] which uses the VAERS dataset to identify COVID-19 vaccine batch IDs with high rates of adverse events. It must be noted that data entry issues with the batch ID in VAERS may generate misleading toxicity information [40]. Lastly, the OpenVAERS site [41] presents "red-box summaries" (**Figure 3a**) and graphs of VAERS database queries (**Figure 3b**), including the frequency of annual death reports before and during the COVID-19 pandemic as well as the timing of reports in days after the injection.

Some COVID-related data have not been provided "voluntarily" but were obtained by community members using freedom-of-information (FOI) requests. The best-known example is a group called Public Health and Medical Professionals for Transparency (PHMPT), whose lawyer Aaron Siri obtained the release of Pfizer's application documents from the US FDA through a court order [42]. In Australia, the Therapeutic Goods Authority maintains a log of FOI releases relating to drug safety dating back to July 2011 [43]. The documents provided are in PDF format, even though some of them include quantitative (tabular) data. The TGA has been criticized for providing only their answers, not the wording of the original requests, which would facilitate understanding the documents provided. As an example, COVID-related documents include the June 2021 release of the 928-page final report on Pfizer's study RN9391R58, which includes some data tables relevant for vaccine safety in pregnant women. The same report, yet with fewer redactions, was released to PHMPT in November 2021.

Figure 3.
Screenshot of OpenVAERS, a website providing moderated access to vaccine adverse event reports as of March 21, 2023. (a) Red-box summaries, and (b) charts of frequency and timing of death reports. Source: https://openvaers.com/covid-data, used with permission.

4. Data-centered COVID-19 narratives and counter-narratives

During the COVID-19 pandemic, many different data-centered narratives emerged that relied on the availability of the near-real time and longer-term open data described in the previous section. The following examples illustrate the use of public health data to amplify fears of the virus in the general population; obfuscate healthcare system usage and overload; coerce people into accepting the COVID-19 vaccines; and mislead the public about the efficacy of face masks. In analogy with the infamous "pandemic of the unvaccinated," we characterize the other narratives as the pandemics of the unafraid, the untreated, and the unmasked, respectively.

4.1 The pandemic of the unafraid

In early March 2023, internal messages by Matt Hancock, the UK's Health Secretary at the beginning of the pandemic, were leaked to the Telegraph newspaper in what should be, by any pre-pandemic standards, a major political scandal. These "Lockdown Files" [44] demonstrate that the UK government deliberately sowed fear in order to secure the public's compliance with pandemic restrictions. Hancock is quoted with WhatsApp messages dated December 13, 2020, discussing "When do we deploy the new variant" and planning to "frighten the pants of everyone with the new strain." The UK along with many other governments appeared to worry that the population was not afraid enough and that a lack of compliance might lead to a "pandemic of the unafraid."

This is perplexing since one of the fundamental rules in public health and outbreak response is to avoid the use of fear narratives [45]. Unfortunately, open data have also been abused to terrorize the population. With reference to epidemic curves of reported or modeled COVID-19 cases, politicians like Ontario's Premier Doug Ford stoked the fear of a "terrible, terrible virus" [46].

The UK's "Lockdown Files" warrant a reminder of the situation in December 2020. Archived BBC News web pages show the pandemic curve for the second wave of

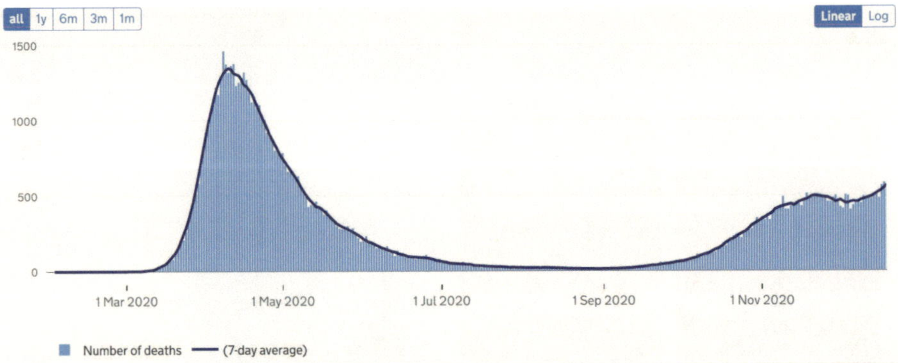

Figure 4.
United Kingdom—Deaths within 28 days of positive test by date of death up to December 19, 2020. Source: https://coronavirus.data.gov.uk/details/deaths?areaType=overview&areaName=United%20Kingdom. Contains public sector information licensed under the Open Government License v3.0.

Covid deaths (based on the government data pictured in **Figure 4**) and an informative footnote stating that "Rules were amended over the summer to include deaths in the coronavirus total only if they occurred within 28 days of a positive test. Previously in England, all deaths after a positive test were included" [47]. Both, the before and after rules almost certainly led to an exaggeration of COVID mortality in public communications, and broader acceptance of the debate about "saving" or "canceling" Christmas 2020. The BBC headlines from the previous day included items like the following [48]:

- Postcode check: Find out the rules where you live.
- Christmas relief: In one city, the urge to meet is greatest of all.
- How schools managed to save the Christmas nativity.
- UK coronavirus cases up by 35,383 on Thursday.
- Canceling Christmas: "They said they understood."
- "Miracle Covid survivor feared his life was over."

Meanwhile, critical research had pointed out the misleading nature of the early 2020 COVID data, maps, charts, and infographics. For example, case counts depended on PCR tests, the use of which for diagnostic testing was critically flawed in several ways [49]. Resulting case counts as well as incidence rates were found to be unsuitable for decision-making [20]. Cartographic issues in online maps were also discussed by several research groups and geospatial industry representatives since February 2020 [6, 7, 50, 51]. Yet, up to the present time, organizations like OWID (**Figure 2**) suggest in the byline of their map and chart titles that the data on cases (wrongly equated with "infections") and deaths due to COVID-19 may be undercounted, without intimating the possibility of the data being overcounted.

4.2 The pandemic of the untreated

While the pandemic of the unafraid reflects a broad-based fear narrative, a concomitant storyline emerged around the overload of hospital capacities early in the pandemic, for example, in the Lombardy region in Italy as well as New York City in the US. This "pandemic of the untreated," or the concern that healthcare systems would be unable to treat everyone equally, was the basis for the initial "two weeks [of lockdown] to flatten the curve" and many subsequent waves of pandemic restrictions. Yet, it was never established that hospitals anywhere in the world were more overloaded during COVID-19 than during other respiratory disease cycles [e.g., [52]]. Intensive-care unit (ICU) capacity and occupancy in Ontario, Canada, hospitals fluctuated throughout the pandemic, as shown in an archived copy of the province's hospitalization status as of February 1, 2022 [53]. There was significant availability (empty beds) at all times and non-COVID occupancy appears to have absorbed much of the episodic COVID-related peaks.

In addition, the impact of the virus itself came under scrutiny when we realized that hospitalizations due to COVID-19 were only a subset of all hospitalizations with a positive COVID-19 test. In January 2022, the Province of Ontario, Canada, added a variable to their hospitalization data to distinguish between COVID-19 as the cause of admission or intensive-care unit treatment on one hand, and COVID-19 as an incidental finding in a patient admitted and/or treated for another, often unrelated condition or injury, on the other hand. For example, the February 1, 2022 report [53] shows 44% "Admitted for other reasons." The government open data charted in **Figure 5** reveals that hospital admissions for other reasons accounted for around 60% of all "COVID-19 hospitalizations" since June 2022. Even among ICU admissions of the past year, between 30% and over 50% were assigned to the COVID-19 statistics although the virus was only an incidental finding. The Canadian media reported about the new variable [e.g., [54]], yet it took many journalists almost three years to acknowledge that "We are overcounting covid deaths and hospitalizations. That's a problem" [55].

4.3 The pandemic of the unvaccinated

The slogan of the "pandemic of the unvaccinated" is attributed to CDC Director Dr. Rochelle Walensky's press briefing on July 16, 2021 [56]. The statement was made with reference to higher proportions of seriously ill COVID patients among unvaccinated Americans than their proportion in the general or elderly populations would suggest. The original phrase referred to the threat to unvaccinated individuals' own health, but it was quickly turned into a debate about the alleged threat posed by unvaccinated individuals and the ethics of prioritizing healthcare services for vaccinated patients [57, 58].

Do open data support the narrative of a pandemic of the unvaccinated? Using the example of the province of Ontario, Canada, during the summer and fall of 2021, the numbers and also the rates of unvaccinated individuals among cases, hospitalizations, and intensive-care unit patients were indeed higher than those of the vaccinated population. With reference to the case rates (**Figure 6a**), it must be noted that the mRNA vaccines were not designed to prevent infection and transmission [59]. Studies conducted during the first phases of rollout showed some reduction in transmission but the higher case rates among unvaccinated individuals in the second half of 2021 could as well be attributed to differences in testing frequency. Since this was a

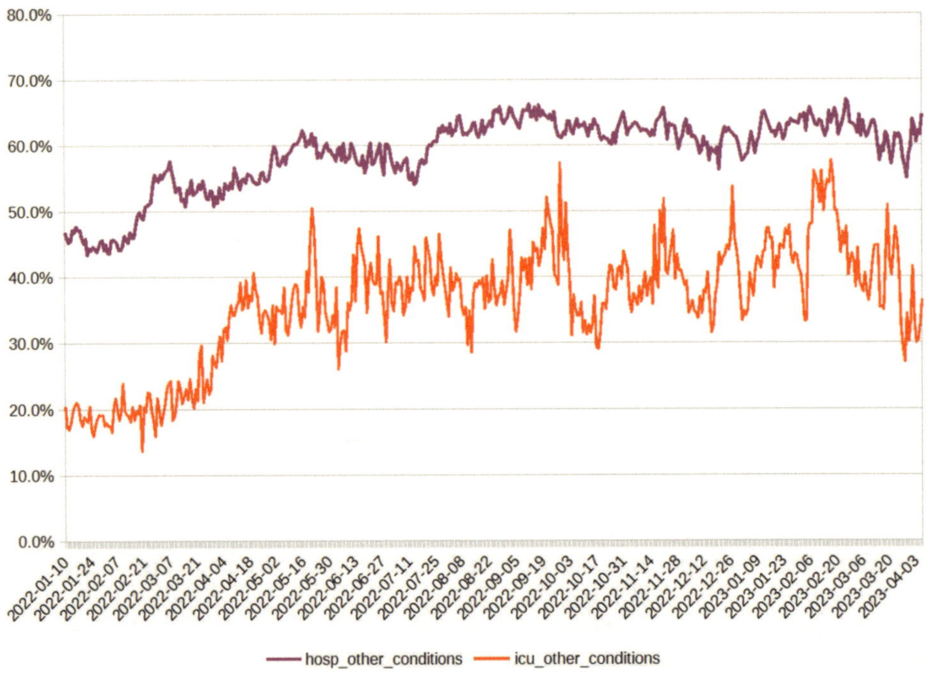

Figure 5.
Proportions of COVID-19 positive hospital and ICU admissions from January 2022 to early April 2023 that occurred for other reasons (not due to COVID). Data source: Ontario Data Catalog, https://data.ontario.ca/en/dataset/breakdown-of-covid-19-positive-hospital-admissions. Contains information licensed under the Open Government License—Ontario.

transition period during the implementation of the vaccine mandates, unvaccinated individuals such as post-secondary students and staff were required to present negative antigen tests to access campuses. With the Omicron variant and wave 5 starting in late 2021 [60], the case rate for vaccinated individuals ("immune escape," "breakthrough" infections) crossed above the unvaccinated rate, and from February to June 2022, the two rates have been similar, while the rate for individuals with a booster dose was noticeably higher in spring 2022 (see **Figure 6a**).

The vaccines were developed to provide protection from symptomatic and severe COVID-19 [59]. This protection was later shown to wane in a span of a few months [61, 62]. Accordingly, the Public Health Ontario data on hospitalizations and ICU occupancy (**Figure 6b**) are higher in unvaccinated patients in 2021 but switch to a higher burden among vaccinated patients in 2022. These data are provided as counts only; rates are not available from this data source. In addition, the agency discontinued the publication of data on hospitalization by vaccination status as of June 30, 2022 with reference to the high fully vaccinated rate of "approximately 87% of eligible Ontarians" [63].

In the UK, independent researchers have highlighted discrepancies in the calculation of vaccination status due to the denominator for the unvaccinated rate [64]. Another example of issues with cross-referencing datasets on hospitalization and vaccination was highlighted in a comment responding to two US-based studies [65]. A newspaper reported in December 2021 in the context of Germany's tightening lockdown rules that the incidence calculation for Bavaria included 57,489 cases with

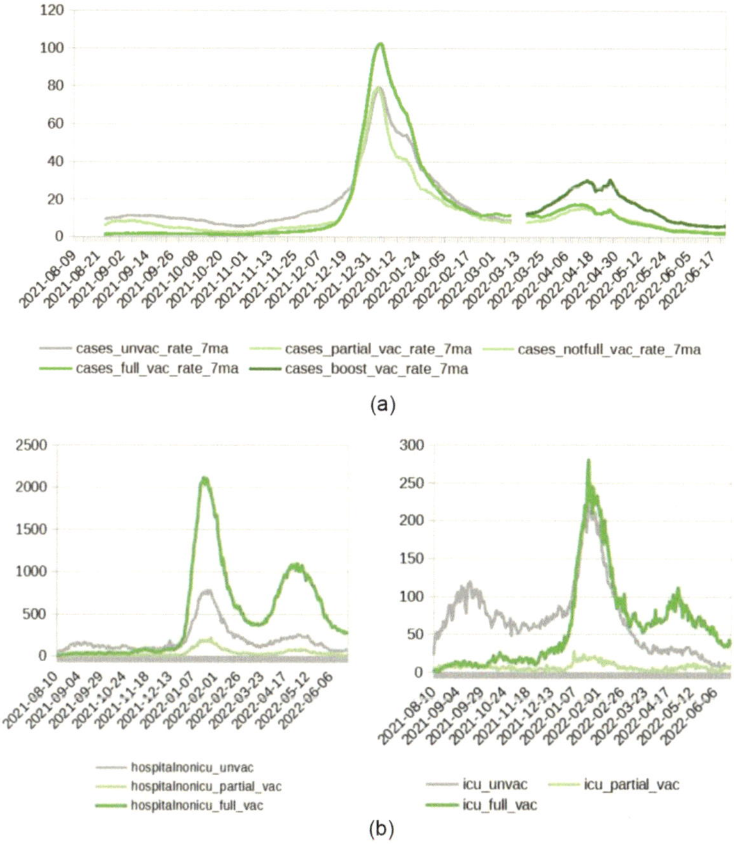

Figure 6.
(a) COVID-19 cases and (b) COVID-19 hospitalization and ICU occupancy rates by vaccination status, August 2021 to June 2022. Data source: Ontario Data Catalog, https://data.ontario.ca/dataset/covid-19-vaccine-data-in-ontario. Contains information licensed under the Open Government License - Ontario.

unknown vaccination status among 72,141 cases counted as unvaccinated [66]; that is, the unvaccinated rate may have been over-estimated almost fivefold due to data inconsistencies and misinterpretation.

4.4 The pandemic of the unmasked

To this day in March 2023, proponents of the COVID fear narratives are holding on to face masks as a tool for community protection. Meanwhile, a number of high-profile studies have shown limited to no evidence of efficacy from different types of face coverings, most recently the updated Cochrane systematic review on "Physical interventions to interrupt or reduce the spread of respiratory viruses" [67]. On the basis of results from multiple randomized-control trials before and during the SARS-CoV-2 pandemic, the authors conclude with moderate certainty that "Wearing masks in the community probably makes little or no difference to the outcome of influenza-like illness (ILI)/COVID-19 like illness compared to not wearing masks" nor to lab-confirmed infection. The confidence intervals for the risk ratios in both groups of trials include 1.0, meaning that there is a possibility that masks increased the risk of infection and/or illness rather than reduce it. In terms of types of face coverings, the review also could not confirm a difference in protection from N95/P2 respirators compared to medical/surgical masks, even in a healthcare setting.

Criticism in the news and social media focused on an overinterpretation of the Cochrane findings. For example, the Washington Post [68] writes "Yet another study on masking causes confusion." Meanwhile their page title reads "The science has not changed. N95 masks still protect against covid," in direct contradiction to the results of the study discussed. The writer in fact responds to misinterpretations of the Cochrane review rather than to the review's findings. She laments that "Some have taken this to mean that masks don't work to protect against the coronavirus," while the review authors only concluded that there is no robust evidence that masks work.

What do the data tell us about community masking? Open data on voluntary masking or compliance with mask mandates are hard to come by, while metrics representing the stringency of government response measures such as the OxGRT index [38] confound masking with other non-pharmaceutical interventions. The most illuminating approach was popularized on social media by Ian Miller, a sports writer and science columnist, and author of "Unmasked: The Global Failure of COVID Mask Mandates" [69]. Miller plots daily or cumulative COVID-19 case rates of different jurisdictions against time and mask-related policies (**Figure 7**). Invariably, the data, which are taken from the WHO, show a respiratory disease cycle that is seemingly independent of public health interventions. While these charts can be viewed as over-simplified and will not replace substantial statistical analyses, they certainly inspire questions about the authoritarian nature of the Western pandemic response [70] and the safety of face masks [71]. Pandemic open data have repeatedly been misused by governments and media, for example, to conjure a "pandemic of the unmasked" during rising infections, in order to push for interventions without proper debate.

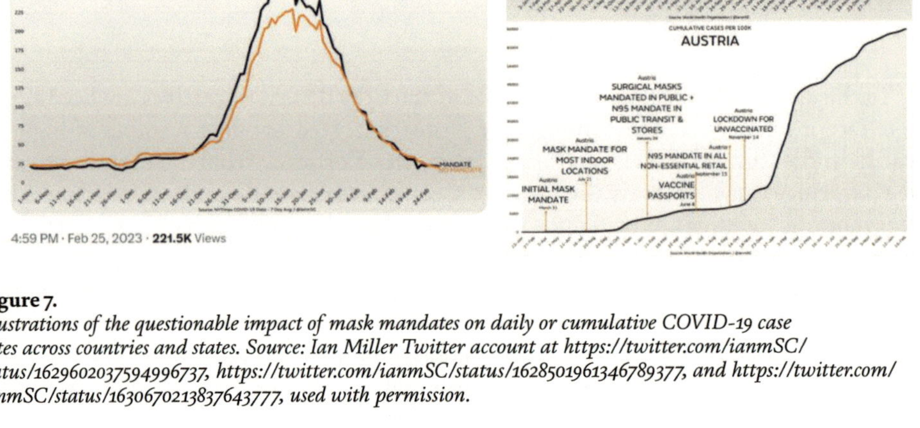

Figure 7.
Illustrations of the questionable impact of mask mandates on daily or cumulative COVID-19 case rates across countries and states. Source: Ian Miller Twitter account at https://twitter.com/ianmSC/status/1629602037594996737, https://twitter.com/ianmSC/status/1628501961346789377, and https://twitter.com/ianmSC/status/1630670213837643777, used with permission.

DOI: http://dx.doi.org/10.5772/intechopen.1001594

5. Conclusion: approaches toward reducing misinformation from, and with, open data

The hierarchical relationship between data, information, and knowledge is often visually represented in the form of a pyramid. At the base of the pyramid are the data: raw and unprocessed facts and figures that are collected across all fields of societal activity. As data are processed, organized, and analyzed, they are transformed into information, thereby becoming more meaningful and useful. As information in turn is understood, applied, and shared, it becomes knowledge, that is, the ability to use information to make decisions, solve problems, or gain insights. If errors are made when processing, organizing, and analyzing the underlying data, misinformation may result, which in turn can lead to wrong decisions. How can we prevent this from happening from, and with, pandemic open data?

It is desirable to have the most disaggregated open data available, subject to the limits of privacy regulations. In public health, individual case data will usually have to be aggregated but the approach used for aggregation is critical. For example, we have seen that vaccinated individuals were usually counted among unvaccinated cases until 14 days after receiving the shots. Instead, there should have been a separate group for those between days 0 and 14 after vaccination, allowing each analyst to decide how to aggregate them, if necessary. When analyzing vaccine efficacy, one might group these "not-yet-protected" with the unvaccinated. However, when analyzing vaccine safety, one would include these "just-vaccinated" with other vaccinated groups.

Equally desirable is transparent metadata. Defining anyone who passed away within 28 days of a positive test as a COVID-19 death is highly unexpected and requires proper labeling as in the above example from the BBC homepage [47]. Explanations from media or public health agencies of the chosen methodologies would go a long way in building trust with the public. For example, in response to a resident's question, Toronto Public Health confirmed that "Individuals who have died with COVID-19, but not as a result of COVID-19 are included in the case counts for COVID-19 deaths in Toronto" [72]. This illustrates that the agencies did not hide their practices, but often failed to explain the reasons behind the skewed statistics and their implications.

Furthermore, survivor bias in comparisons between categories such as unvaccinated and vaccinated needs to be addressed. Due to the one-directional movement of individuals from one category to the other (but never back), the time spent in the first category must be taken into account by using person-time as a unit rather than simple person-count [73].

When public health agencies and many academic scientists are captured, we depend on investigative journalists and citizen researchers to challenge state-controlled narratives. Instead, "many historic norms of journalism were ditched in newsrooms around the world" [74], norms regarding critical unbiased investigations and speaking truth to power. Additionally, "laypersons" were told by the media, "You Must Not 'Do Your Own Research' When It Comes To Science" [75]. In such an environment of one-sided reporting and lack of a diversity of allowable opinions, open data become an essential resource for a growing and striving citizen research ecosystem. The openness of COVID-19 data, therefore, was a blessing, if in disguise.

Acknowledgements

This research was partially funded through a grant from the Natural Sciences and Engineering Research Council of Canada.

Author details

Claus Rinner
Toronto Metropolitan University, Toronto, Canada

*Address all correspondence to: crinner@torontomu.ca

References

[1] Open Knowledge Foundation. The Open Definition, Available from: http://opendefinition.org/

[2] Open Data Charter. Principles. Available from: https://opendatacharter.net/principles/

[3] Fast V, Rinner C. Mediating open data: Providers, portals, and platforms. Editorial, Journal of the Urban and Regional Information Systems Association. 2017;**28**(1):7-8

[4] Baculi E, Fast V, Rinner C. The geospatial contents of municipal and regional open data catalogs in Canada. Journal of the Urban and Regional Information Systems Association. 2017;**28**(1):39-48

[5] Greene S, Rinner C. Examining the value of geospatial open data. In: Robinson PJ, Scassa T, editors. The Future of Open Data. Ottawa, ON: University of Ottawa Press. pp. 159-178

[6] Mooney P, Juhász L. Mapping COVID-19: How web-based maps contribute to the infodemic. Dialogues in Human Geography. 2020;**10**(2):265-270

[7] Rinner C. Mapping COVID-19 in context: Promoting a proportionate perspective on the pandemic. Cartographica. 2021;**56**:14-26

[8] Ioannidis JPA. A Fiasco in the Making? As the Coronavirus Pandemic Takes Hold, We Are Making Decisions Without Reliable Data. Boston, MA: Stat News; 2020. Available from: https://www.statnews.com/2020/03/17/a-fiasco-in-the-making-as-the-coronavirus-pandemic-takes-hold-we-are-making-decisions-without-reliable-data/

[9] Boccia S, Ricciardi W, Ioannidis JPA. What other countries can learn from Italy during the COVID-19 pandemic. JAMA Internal Medicine. 2020;**180**:927-928

[10] Chin V, Samia NI, Marchant R, Rosen O, Ioannidis JPA, Tanner MA, et al. A case study in model failure? COVID-19 daily deaths and ICU bed utilisation predictions in New York state. Journal European Journal of Epidemiology. 2020;**35**:733-742

[11] Bendavid E, Mulaney B, Sood N, Shah S, Bromley-Dulfano R, Lai C, et al. Covid-19 antibody seroprevalence in Santa Clara county, California. International Journal of Epidemiology. 2021;**50**:410-419

[12] Ioannidis JPA. Infection fatality rate of COVID-19 inferred from seroprevalence data. Bulletin of the World Health Organization. 2021;**99**:19-33F

[13] Ioannidis JPA. Coronavirus disease 2019: The harms of exaggerated information and non-evidence-based measures. European Journal of Clinical Investigation. 2020;**50**(4):e13222

[14] Jefferson T, Heneghan C, Spencer E, Brassey J. Are You Infectious if you have a Positive PCR Test Result for COVID-19. Oxford, UK: Centre for Evidence-Based Medicine, University of Oxford; 2020. Available from: https://www.cebm.net/covid-19/infectious-positive-pcr-test-result-covid-19/

[15] Heneghan C, Oke J. Public Health England has Changed its Definition of Deaths: Here's what it Means. Oxford, UK: Centre for Evidence-Based Medicine, University of Oxford; 2020.

Available from: https://www.cebm.net/covid-19/public-health-england-death-data-revised/

[16] Fenton NE. The Bangladesh Mask Study: A Bayesian Perspective. Preprint. 2022. Available from: https://www.researchgate.net/publication/360320982

[17] McLachlan S, Osman M, Dube K, Chiketero P, Choi Y, Fenton N. Analysis of COVID-19 Vaccine Death Reports from the Vaccine Adverse Events Reporting System (VAERS) Database. Preprint. 2021. Available from: https://www.researchgate.net/publication/352837543

[18] Fenton N, Neil M. Flawed Covid definitions, data and modelling. Blog post, "Where are the numbers?" on Substack. 2022. Available from: https://wherearethenumbers.substack.com/p/flawed-covid-definitions-data-and

[19] Kuhbandner C, Homburg S, Walach H, Hockertz S. Was Germany's Lockdown in Spring 2020 necessary? How bad data quality can turn a simulation into a delusion that shapes the future. Futures. 2022;**135**:102879

[20] Hirsch O, Bergholz W, Kisielinski K, Giboni P, Sönnichsen A. Methodological problems of SARS-CoV-2 rapid point-of-care tests when used in mass testing. AIMS Public Health. 2022;**9**(1):73-93

[21] Thacker PD. Covid-19: Researcher blows the whistle on data integrity issues in Pfizer's vaccine trial. British Medical Journal. 2021;**375**:n2635

[22] Shir-Raz Y, Elisha E, Martin B, Ronel N, Guetzkow J. Censorship and suppression of Covid-19 Heterodoxy: Tactics and counter-tactics. Minerva. 2022. DOI: 10.1007/s11024-022-09479-4 [Published online ahead of print]

[23] Ioannidis JPA. How the Pandemic is Changing the Norms of Science. Tablet Magazine. 2021. Available from: https://www.tabletmag.com/sections/science/articles/pandemic-science

[24] World Health Organization. WHO Coronavirus (COVID-19) Dashboard, Available from: https://covid19.who.int/

[25] European Centre for Disease Prevention and Control. Latest COVID-19 Data. Available from: https://www.ecdc.europa.eu/en/covid-19/data

[26] Centers for Disease Control and Prevention. COVID Data Tracker. Atlanta, GA: US Department of Health and Human Services, CDC; 2023 Available from: https://covid.cdc.gov/covid-data-tracker

[27] Public Health Agency of Canada. Coronavirus Disease (COVID-19). Available from: https://www.canada.ca/en/public-health/services/diseases/coronavirus-disease-covid-19.html

[28] Office for National Statistics. Coronavirus (COVID-19). Available from: https://www.ons.gov.uk/peoplepopulationandcommunity/healthandsocialcare/conditionsanddiseases

[29] Robert Koch-Institut. Coronavirus SARS-CoV-2. Available from: https://www.rki.de/DE/Content/InfAZ/N/Neuartiges_Coronavirus/nCoV_node.html

[30] Public Health Ontario. COVID-19 in Ontario: Focus on May 22, 2022 to May 28, 2022. Weekly Epidemiological Summary (Archived). Available from: https://www.publichealthontario.ca/-/media/Documents/nCoV/epi/2022/05/covid-19-weekly-epi-summary-report-may-28.pdf

[31] Centers for Disease Control and Prevention. Vaccine Safety, Vaccine

Adverse Event Reporting System (VAERS). Available from: https://www.cdc.gov/vaccinesafety/ensuringsafety/monitoring/vaers/index.html

[32] Worldometer. COVID – Coronavirus Statistics. Available from: https://www.worldometers.info/coronavirus/

[33] Mathieu E, Ritchie H, Rodés-Guirao L, Appel C, Giattino C, Hasell J, Macdonald B, Dattani S, Beltekian D, Ortiz-Ospina E, Roser M. Coronavirus Pandemic (COVID-19). 2020. Available from: OurWorldInData.org, https://ourworldindata.org/coronavirus

[34] Center for Systems Science and Engineering. COVID-19 Dashboard. Baltimore, MD: Johns Hopkins University. Available from: https://coronavirus.jhu.edu/map.html

[35] Dong E, Du H, Gardner L. An interactive web-based dashboard to track COVID-19 in real time. The Lancet Infectious Diseases. 2020;**20**(5):533-534. DOI: 10.1016/S1473-3099(20)30120-1

[36] The COVID Tracking Project. The Atlantic. Available from: https://covidtracking.com/

[37] Esri Canada Inc. COVID-19 Canada, COVID-19 Resources. Available from: https://resources-covid19canada.hub.arcgis.com/

[38] Hale T, Angrist N, Goldszmidt R, Kira B, Petherick A, Phillips T, et al. A global panel database of pandemic policies (Oxford COVID-19 Government Response Tracker). Nature Human Behaviour. 2021;**5**(4):529-538. DOI: 10.1038/s41562-021-01079-8

[39] How Bad is My Batch. Available from: https://howbadismybatch.com/

[40] Ontario Civil Liberties Association. OCLA Statement on Analysis of Batch-Specific Toxicity of COVID-19 Vaccine Products using VAERS Data. Available from: https://ocla.ca/ocla-statement-on-analysis-of-batch-specific-toxicity-of-covid-19-vaccine-products-using-vaers-data/

[41] OpenVAERS. VAERS COVID Vaccine Adverse Event Reports. Available from: https://openvaers.com/covid-data

[42] Public Health and Medical Professionals for Transparency. Pfizer's Documents. Online Document Archive. Available from: https://phmpt.org/pfizers-documents/

[43] Therapeutic Goods Auhority. FOI disclosure log. Online Document Archive. Available from: https://www.tga.gov.au/foi-disclosure-log

[44] The Telegraph. The Lockdown Files. Investigative Report. Available from: https://www.telegraph.co.uk/news/lockdown-files/

[45] Stolow JA, Moses LM, Lederer AM, Carter R. How fear appeal approaches in COVID-19 health communication may be harming the global community. Health Education & Behavior. 2020;**47**(4):531-535

[46] CTV News. Ontario Economy to Reopen 'in a Trickle' Premier Says, After Suggesting Some Relaxation by Victoria Day. 2020. Available from: https://ottawa.ctvnews.ca/ontario-economy-to-reopen-in-a-trickle-premier-says-after-suggesting-some-relaxation-by-victoria-day-1.4906840

[47] BBC News. 2020. Available from: https://web.archive.org/web/20201219034027/https://www.bbc.com/news/uk-51768274

[48] BBC News. 2020. Available from: https://web.archive.org/

web/20201218100523/https://www.bbc.com/news/coronavirus

[49] Tang Y-W, Schmitz JE, Persing DH, Stratton CW. Laboratory diagnosis of COVID-19: Current issues and challenges. Journal of Clinical Microbiology. 2020;**58**(6):e00512-e00520

[50] Field K. Mapping Coronavirus, Responsibly. Blog Post. 2020. Available from: https://www.esri.com/arcgis-blog/products/product/mapping/mapping-coronavirus-responsibly/

[51] Juergens C. Trustworthy COVID-19 mapping: Geo-spatial data literacy aspects of choropleth maps. KN - Journal of Cartography and Geographic Information. 2020;**70**:155-161

[52] National Health Service (NHS) in England. Bed Availability and Occupancy – Overnight. Available from: https://www.england.nhs.uk/statistics/statistical-work-areas/bed-availability-and-occupancy/bed-data-overnight/

[53] Government of Ontario. Hospitalizations | COVID-19 (coronavirus) in Ontario. 2022. Available from: https://web.archive.org/web/20220201215018/https://covid-19.ontario.ca/data/hospitalizations

[54] Global News. 46% of those currently hospitalized with COVID in Ontario were admitted for other reasons: New data. By Ryan Rocca. 2022. Available from: https://globalnews.ca/news/8502714/ontario-incidental-covid-hospitalizations-january-11/

[55] Washington Post. We are Overcounting Covid Deaths and Hospitalizations. That's a Problem. By Leana S. Wen, Contributing columnist. 2023. Available from: https://www.washingtonpost.com/opinions/2023/01/13/covid-pandemic-deaths-hospitalizations-overcounting/

[56] The New York Times. C.D.C. Director Warns of a 'Pandemic of the Unvaccinated'. By Emily Anthes and Alexandra E. Petri. 2021. Available from: https://www.nytimes.com/2021/07/16/health/covid-delta-cdc-walensky.html

[57] Goldman E. How the unvaccinated threaten the vaccinated for COVID-19: A Darwinian perspective. Proceedings of the National Academy of Sciences of the United States of America. 2021;**118**(39):e2114279118

[58] Kampf G. COVID-19: Stigmatising the unvaccinated is not justified. The Lancet. 2021;**398**(10314):1871

[59] Doshi P. Will covid-19 vaccines save lives? Current trials aren't designed to tell us. BMJ. 2020;**371**:m4037

[60] Public Health Agency of Canada. Federal, Provincial, Territorial Public Health Response Plan for Ongoing Management of COVID-19, 3rd edition. Available from: https://www.canada.ca/en/public-health/services/diseases/2019-novel-coronavirus-infection/guidance-documents/federal-provincial-territorial-public-health-response-plan-ongoing-management-covid-19.html

[61] Tartof SY et al. Effectiveness of mRNA BNT162b2 COVID-19 vaccine up to 6 months in a large integrated health system in the USA: A retrospective cohort study. The Lancet. 2021;**398**(10309):1407-1416

[62] Ferdinands JM et al. Waning of vaccine effectiveness against moderate and severe covid-19 among adults in the US from the VISION network: test negative, case-control study. BMJ. 2022;**379**:e072141

[63] Government of Ontario. Data Catalogue, Hospitalizations by Vaccination Status. Available from: https://data2.ontario.ca/en/dataset/

covid-19-vaccine-data-in-ontario/resource/274b819c-5d69-4539-a4db-f2950794138c

[64] Neil M, Fenton N, Smalley J, Craig C, Guetzkow J, McLachlan S, Rose J. Official Mortality Data for England Suggest Systematic Miscategorisation of Vaccine Status and Uncertain Effectiveness of Covid-19 Vaccination. 2022. Available from: https://www.researchgate.net/profile/Martin-Neil-2/publication/357778435

[65] Yim P. Concerns about COVID-19 vaccination observational trials in the United States. JAMA Internal Medicine. 2022;**182**(10):1071-1081. Available from: https://jamanetwork.com/journals/jamainternalmedicine/fullarticle/2796235

[66] Berliner Zeitung. Verzerrung der Statistik? Bayern zählt unbekannten Impfstatus als ungeimpft. [Skewed statistics? Bavaria counts unknown vaccination status as unvaccinated.]. 2021. Available from: https://www.berliner-zeitung.de/news/verzerrung-der-statistik-bayern-zaehlt-unbekannten-impfstatus-als-ungeimpft-li.199206

[67] Jefferson T et al. Physical Interventions to Interrupt or Reduce the Spread of Respiratory Viruses. Cochrane Database of Systematic Reviews. 2023. Available from: https://www.cochranelibrary.com/cdsr/doi/10.1002/14651858.CD006207.pub6/full

[68] Washington Post. Yet Another Study on Masking Causes Confusion. By Leana S. Wen, Contributing Columnist. 2023. Available from: https://www.washingtonpost.com/opinions/2023/02/16/cochrane-study-masks-covid-pandemic/

[69] Miller I. Unmasked: The Global Failure of COVID Mask Mandates. Brentwood, TN: Post Hill Press. p. 232

[70] Simandan D, Rinner C, Capurri V. The academic left, human geography, and the rise of authoritarianism during the COVID-19 pandemic. Geografiska Annaler B. Human Geography. Published online: 23 January 2023. DOI: 10.1080/04353684.2023.2168560

[71] Kisielinski K, Giboni P, Prescher A, Klosterhalfen B, Graessel D, Funken S, et al. Is a mask that covers the mouth and nose free from undesirable side effects in everyday use and free of potential hazards? International Journal of Environmental Research and Public Health. 2021;**18**(8):4344

[72] Toronto Public Health. Individuals Who Have Died With COVID-19, but Not as a Result of COVID-19 are Included in the Case Counts for COVID-19 Deaths in Toronto. Tweet. 2020. Available from: https://twitter.com/TOPublicHealth/status/1275888390060285967

[73] Fenton N. 'Never Vaccinated' vs 'Ever Vaccinated' Mortality Rate Illusion. Blog Post. 2023. Available from: https://wherearethenumbers.substack.com/p/never-vaccinated-vs-ever-vaccinated

[74] Trish Wood. CBC Journalist Quits over Biased COVID Coverage and I Get Cancelled (Temporarily) for Telling Her Story. Blog Post and Podcast. 2022. Available from: https://trishwood.substack.com/p/cbc-journalist-quits-over-biased

[75] Forbes. You Must Not 'do Your Own Research' When it Comes to Science. By Ethan Siegel. 2020. Available from: https://www.forbes.com/sites/startswithabang/2020/07/30/you-must-not-do-your-own-research-when-it-comes-to-science/

Chapter 3

Trends in High-Performance Data Engineering for Data Analytics

Vibhatha Abeykoon and Geoffrey Charles Fox

Abstract

Over the past decade, data analytics has undergone significant transformation due to the increasing availability of data and the need to extract valuable insights from it. However, the classical big data stack needs to be faster in data engineering, highlighting the need for high-performance computing. Data analytics has motivated the engineering community to build diverse frameworks, including Apache Arrow, Apache Parquet, Twister2, Cylon, Velox, and Datafusion. These frameworks have been designed to provide high-performance data processing on C++-backed core APIs, with extended usability through support for Python and R. Our research focuses on the trends in the evolution of data engineering, which have been characterized by a tendency towards high-performance computing, with frameworks designed to keep up with the evolving demands of the field. Our findings show that the modern-day data analytics frameworks have been developed with C++ core compute and communication kernels and are designed to facilitate high-performance data processing. And this has been a critical motivation to develop scalable components for data engineering frameworks.

Keywords: data engineering, data analytics, high-performance computing, big-data, data-engineering trends

1. Introduction

In today's data-driven world, data analytics has become essential for businesses and organizations to extract insights from large volumes of data. However, as data grows, traditional data processing methods have proven inadequate, highlighting the need for high-performance data engineering. High-performance data engineering is a process that involves the efficient processing of large volumes of data using advanced computing systems and state-of-the-art hardware. The field of high-performance data engineering for data analytics has been evolving rapidly in recent years, driven by the need for faster, more efficient ways of processing data. This has led to the development of a range of frameworks, tools, and technologies that are designed to facilitate high-performance data processing on a large scale. In our study, we observe that the evolution in data engineering directly impacts data analytics frameworks. Sequential or multi-threaded approaches to computation cannot help analyze more extensive datasets. In replacing parameter servers, data analytics frameworks have evolved to use high-performance computing concepts to build the core compute and communication kernels. Driven by this fact, modern-day data engineering has developed in many ways. Our study mainly focuses on a few aspects of the evolution of data engineering.

The first key aspect is re-writing existing big-data stack with high-performance computing concepts. Language preference has changed from using Java to C++ when it comes to modern-day data engineering frameworks. The second key aspect is the shift in communication models. Rather than using master-slave communication architecture, more platforms have introduced bulk-synchronous parallel data processing to gain more performance. In addition, some frameworks have a preference for remote procedure call-based approaches. The third and final key aspect is the usability and extensibility of data analytics workloads. In developing data analytics algorithms for production, a development stage is defined as an exploratory data analytics stage. In this stage, it is vital that the data scientist can run various feature engineering algorithms and extract features to try new ideas. Unlike classical Java-based development approaches, modern data engineering frameworks have shown signs of extending their usability to Python and R. Still; they retain performance, unlike classical big-data systems having similar Python or R bindings.

Furthermore, data processing and planning have evolved in a new direction where users can now define a plan to run their queries with various platforms without introducing new code. The idea is that a new specification will determine what happens to the data, and the query processing engines that adopt this specification can trivially run the workloads. In Section 2, we discuss how data analytics has evolved in supporting AlexNet [1] to ChatGPT [2]. Section 3 discusses how data engineering has evolved from Apache Hadoop [3] to Ray [4] and how the query processing technologies have adopted new standards to run queries in various query engines with minimal effort. In Section 4, we summarize how data analytics has impacted the data engineering evolution in the past decade.

2. Modern data analytics and practices

Since AlexNet [1] took over the data analytics world by surprise, the data analytics world required more tools to be built on, making deep learning concepts available to build intelligent applications. Before AlexNet, it support vector machines which were breaking records in classifying and recognizing details intelligently. The main difference between these two algorithms is that AlexNet contains more layers of computation done upon the output of a previous layer. This required way more computing power compared to what SVM needed. Also, unlike SVM, AlexNet was not designed to run on CPUs but on GPUs. This is the first time the computer graphics card has been used to do something new.

With the spark of this new idea, image classification algorithms evolved. Writing complex neural networks requires more computing capability, data, and platforms. Existing tooling in the early days of deep learning required users to write computation layers using CUDA compute kernels from scratch, and this is not a very scalable idea. Unifying various concepts in neural networks, libraries sprang up to provide better tooling to develop deep learning applications.

PyTorch, Tensorflow, Apache MxNet, Caffe, Theano, and Chainer are some of the most prominent deep-learning libraries. These frameworks can provide higher-level

APIs to build neural networks in a short time. Underneath these frameworks, use accelerators like GPUs and TPUs perform much better than doing computation on CPUs. But the user did not have to worry about the computation APIs; instead, they just had to select which accelerator they wanted to use when developing applications. Regarding usability, Python was the go-to language for writing such analytical applications. For the moment, we will not go into a deeper analysis of why Python was selected. Still, Python was already a popular language, and it was easy to use because it did not need pre-compilation.

2.1 Data analytics frameworks

So far, we have discussed how data analytics have evolved and the role of data analytics frameworks. It is better to understand this deeply by looking into each framework. Among the data analytics frameworks, legacy machine learning frameworks like Scikit-learn [5] contained many statistical and machine learning models. But before moving into details of the most promising systems, it is worth mentioning that deep learning-based solutions have evolved way beyond the primary machine leanring models.

- PyTorch [6]: An open-source machine learning library based on Torch, which is used for applications such as natural language processing and computer vision. PyTorch is known for its dynamic computational graph, which allows for more flexibility in model creation and training. It is widely used in both academia and industry.

- TensorFlow [7]: A popular open-source framework for machine learning and deep learning that was developed by Google Brain. It supports a wide range of tasks, including image and speech recognition, and has a large and active community of users and contributors. TensorFlow's key features include its data flow graph architecture and its ability to scale across multiple devices.

- MXNet [8]: A flexible and efficient deep learning framework that is known for its fast training speeds and low memory usage. Developed by Amazon Web Services, MXNet supports multiple programming languages and has a variety of pre-built models for image, text, and speech recognition.

- Caffe [9]: A deep learning framework that is focused on speed and scalability. It was developed by the Berkeley Vision and Learning Center and is known for its ease of use and powerful visualization tools. Caffe is commonly used for image classification, segmentation, and object detection.

- Theano [10]: A Python library that allows for efficient computation of mathematical expressions, particularly in the context of deep learning. It is known for its ability to optimize CPU and GPU usage and for its strong integration with NumPy. Theano is used in a variety of applications, including natural language processing and computer vision.

- Chainer [11]: A Python-based deep learning framework that was developed by Japanese company Preferred Networks. Chainer is known for its dynamic computational graph, which allows for more flexibility in model creation and training. It supports a wide range of tasks, including image and speech recognition, and has a variety of built-in optimization algorithms.

Even though frameworks like Tensorflow, MxNet and Theano are used by many data analysts and industrial work, it is worth noting that PyTorch has become the de-facto standard in the research community to develop deep learning models. The primary reason is it's a configurable and wide array of APIs to break down the application development into finer details.

PyTorch is a machine-learning library that is built on top of the Torch library. It is designed to provide a user-friendly and flexible framework for building and training deep neural networks in Python. PyTorch offers support for dynamic computation graphs, which allows for modifying neural network architecture during runtime. PyTorch supports CPU, GPU, and TPU-based computation models, which are mainly designed to run on a bulk-synchronous-parallel communication model. In simple terms, AllReduce collective is regarded as the widely used communication operation when running distributed training programmes. In addition, PyTorch recently introduced an RPC-based computation model to provide more flexibility in running applications in cloud environments.

A wide variety of accelerators and communication models enables researchers to experiment with different model architectures and ideas without committing to a fixed architecture beforehand. It offers many built-in functions and classes for building and training neural networks, including modules for convolutional and recurrent neural networks, optimization algorithms, and loss functions. It also supports GPU acceleration, which enables efficient movement of deep neural networks on modern hardware. PyTorch has gained popularity in the machine learning community due to its ease of use, flexibility, and powerful capabilities. It is widely used in research and industry for various applications such as computer vision, natural language processing, speech recognition, and reinforcement learning.

2.2 Machine learning in production

Python provided an easier way to prototype applications, and later in production, these applications can be compiled into high-performance scripts using tools like Torchscript [12]. By decoupling the model from any runtime environment, TorchScript enables the model to be executed independently of the framework or platform it was developed on. This eliminates the Global Interpreter Lock (GIL) in Python, which can be a bottleneck for executing multithreaded inference. TorchScript prioritizes optimizing the entire program as a whole rather than just individual components or parts.

2.3 Computation intensity in data analytics problems

The performance required to run the most recent data analytics problems has exponentially evolved through the past decade. From image classifiers to intuitive Chess Players to intelligent chatbots, the cost of learning more has risen to new levels. **Figure 1** [13] shows the computing cost for various deep learning models over time.

The data required to get an accurate model would depend on the model. Still, the modern-day argument in data-centric artificial intelligence [14] is that it is better to focus on a data-driven approach than a model-driven approach where the model is

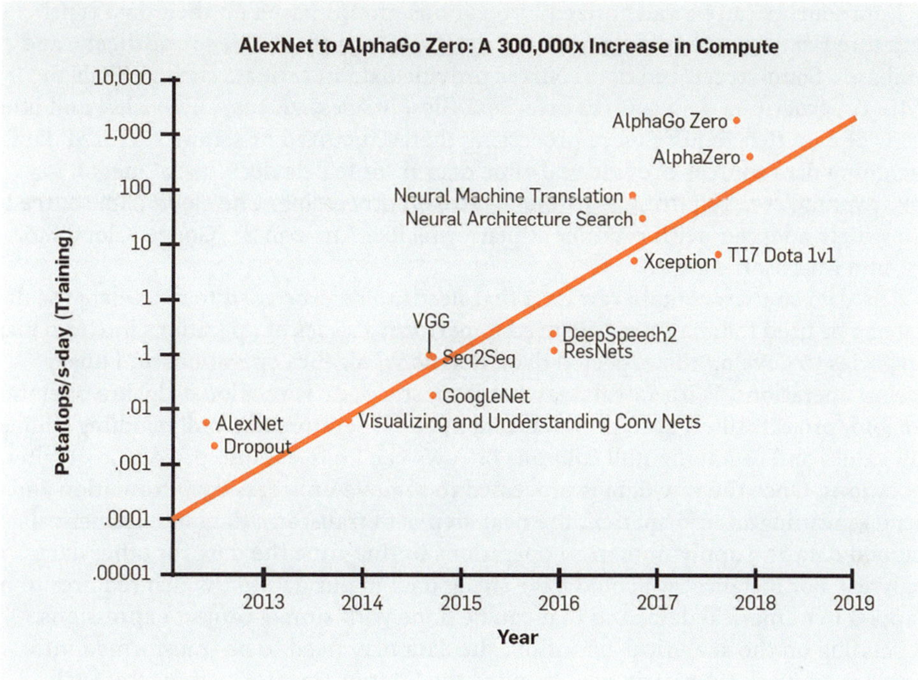

Figure 1.
Amount of compute used in deep learning.

tuned to fit the data. This means that more and more data is required to get a better understanding, requiring data pre-processing at a larger scale.

Can we rely on the current data processing stack or enhance it? This is a significant question that needs to be addressed. How should the systems evolve? What are the best tools, or do we have to write new tools? How should we write such systems? All these questions are valid and very important. Let us learn more about data processing and how such systems have evolved in the past decade.

3. Data engineering

Data Engineering is a discipline that focuses on the design, construction, and maintenance of systems and processes to manage, store, and extract value from large and complex data sets. Breaking down data engineering would give us the following steps;

- Read raw data from data sources
- Formulate a series of operations to process data
- Convert process data to expected output format
- Persistent storage or sending information to a different service
- Fault-tolerant for any operation executed with a defined granularity

Data sources can be categorized into various groups based on their data type. Structured data sources offer organized data, for example, CSV, spreadsheets, and databases. Semi-structured data sources provide data in formats such as JSON or XML. Unstructured data sources offer text files, images, videos, audio files, and other forms of data that require more processing than structured or semi-structured data. Streaming data sources provide real-time data from IoT devices, social media, log files, gaming, etc., requiring real-time or stream processing. The cloud data source is also widely adopted, with prominent platforms like Amazon S3, Google Cloud Storage, and Microsoft Azure.

The data sources contain raw data that needs to be processed to formulate the data that can be used for analytics. We need to perform a series of operations into two main categories to develop the expected data. Relational algebra operations and linear algebra operations. With raw data, what is mostly done is relational algebra operators like join, project, filter, sort, product, union, etc. Operators like null handling, filling null values and removing null columns or rows can be fused into projection or filter operations. Once the raw data is processed to remove unnecessary information and extract meaningful information, the next step is to transform data into numerical mapped data and apply numerical operations to fine-tune the data for other data analytics. For instance, we could have string data in our dataset, which requires to be mapped to numerical data, and that can be done with simple project expressions. Depending on the analytical algorithm, the data may need to be transformed into matrices, normalized matrix data, applied the Fourier transformation, etc. Such operations fall into the linear algebra operations.

Once the data is processed, the output data needs to be logged or fed into other systems in various formats. Some systems expect the data in specific storage formats like Parquet, CSV, HDF5, etc. Thus the processed data also have a data format. Converting data into a different format within systems can cause reading all the data, at least in chunks, and it will cost a lot to access storage and computing power. So the correct format is chosen at the data cleaning/ processing stage, and data is stored. This approach will benefit smaller datasets requiring less system memory and storage. With efficient networking capability on Infini-band, Cray Aries, Intel Omni-Path and similar technologies, data could be moved from the data processing stage to the data analytics stage over in-memory data formats like Apache Arrow [15] in a very efficient way. For instance, Apache Arrow IPC format enables persistent storage and efficient data reading. With Apache Arrow Flight [16], data can be efficiently transferred for remote processing.

We have discussed the main steps of processing and storing data in various stages of the data processing pipeline. But fault tolerance is an essential feature to ensure systems run end-to-end seamlessly. As the term suggests, a system should be able to withstand faults. In the data processing context, such faults can be server shutdowns, network failures, out-of-memory exceptions and runtime exceptions due to unexpected failures in connected services that could stop the system and bring it to a complete stop. If the data processing time is minimal, re-running the workflow from the start point will be relatively inexpensive. However, executing operations like joins could be costly when the dataset is extensive. Computing and storage are not free, so we must repeatedly pay for the resources each time the system fails. To avoid such issues, systems are designed to snapshot certain stages as checkpoints, resume from the most recent checkpoint, and carry out the task.

3.1 Data engineering frameworks

Large datasets cannot be efficiently processed with only a set of relational and linear algebra operators. As the amount of data generated every day continues to increase exponentially, reaching petabytes of data across various platforms serving millions of users, there is a need for a more versatile set of operators and operation modes. Many applications require real-time notifications, forecasts, instant messaging, and reporting within a specific timeframe. To support these requirements, data engineering frameworks encapsulate ways such as stream data processing and batch data processing. These frameworks provide the necessary infrastructure to process large datasets efficiently and effectively, making extracting valuable insights and information easier.

Such widely used data engineering frameworks are;

- Apache Hadoop: An open-source big data framework that provides distributed storage and processing of large data sets using a cluster of commodity hardware. Hadoop is used for batch processing, data analysis, machine learning, and more.

- Apache Spark [17]: A unified analytics engine for large-scale data processing that supports batch processing, streaming, and machine learning. Spark is designed to be fast and efficient, with in-memory computing and the ability to process data in parallel.

- Apache Kafka [18]: A distributed streaming platform that allows users to publish and subscribe to streams of records. Kafka is used for real-time data processing, data streaming, and data integration across different systems.

- Apache Flink [19]: A distributed stream processing engine that supports real-time stream processing and batch processing. Flink is designed to be highly scalable, fault-tolerant, and efficient, with support for both batch and stream processing in one system.

- Apache Beam [20]: An open-source, unified programming model for batch and streaming data processing. Beam provides a simple, consistent API for building data processing pipelines that can run on multiple processing engines, such as Apache Flink, Apache Spark, and Google Cloud Dataflow.

- Apache Storm [21]: a distributed stream processing framework that supports real-time processing of high-velocity data streams. Storm is used for real-time analytics, machine learning, and other applications that require fast and reliable data processing.

- Google Cloud Dataflow [22]: A fully managed, serverless data processing service that enables users to build batch and streaming data pipelines using Apache Beam programming model.

These frameworks are designed to solve a set of problems emerged through out the history of data processing. **Table 1** shows the breakdown of pros and cons in these frameworks based on qualitative features like scalability, fault-tolerance and processing modes.

Framework	Batch Processing	Stream Processing	Fault Tolerance	Programming Model [a]
Apache Hadoop	Yes	No	Yes	MapReduce
Apache Spark	Yes	Yes	Yes	DataFrame
Apache Kafka	No	Yes	Yes	Pub/Sub
Apache Flink	Yes	Yes	Yes	Dataflow
Apache Beam	Yes	Yes	Yes	Unified
Apache Storm	No	Yes	Yes	Spout/Bolt
Google Cloud Dataflow	Yes	Yes	Yes	Unified

[a] *Programming model refers to the type of programming model or approach that each data processing framework uses.*

Table 1.
Features of various data processing frameworks.

Apache Hadoop can be recognized as one of the earliest open-source big-data systems, and with time each system added its unique set of features. Apache software foundation and open-source software development significantly impact improving these systems. To see the progress made by these frameworks, we can look at **Figure 2**, which depicts the Github statistics.

Although these are the widely adopted data processing systems, researchers and engineers have made an enormous shift to gain much better performance. We start to see a pattern when looking into the programming languages used to develop the frameworks mentioned below (**Table 2**).

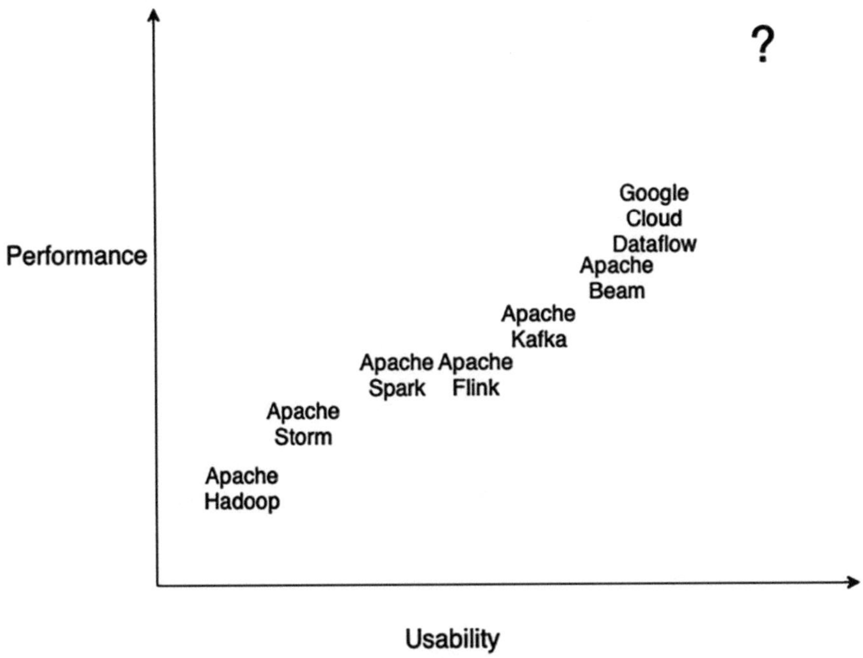

Figure 2.
Github usage statistics of big data frameworks.

Programming language	Frameworks
Java	Apache Hadoop, Apache Flink, Apache Storm
Scala	Apache Spark
Java/Scala	Apache Kafka, Apache Beam
Python	Google Cloud Dataflow

Table 2.
Frameworks grouped by programming language.

3.2 Data engineering evolution

Java and Scala are the programming languages used in developing these big-data systems. Even though Java is known for platform independence, security, and scalability, the performance aspect could be more pleasing for applications associated with heavy computing tasks. In that aspect, C++ or Rust are better replacements to gain much better performance.

Using high-performance languages like C++ and Rust has been one of the trends in modern large-scale data processing systems. After the big-data era, frameworks like Ray, Velox, DataFusion and Apache Arrow have been mostly used to design data processing pipelines. One key takeaway from these frameworks is how they provide usability. Unlike classical, big data systems, which focused more on languages like Scala and Java to give the majority of their APIs, these frameworks focused on the Python programming language to a greater extent. Big-data systems provided Python wrappers known to slow performance due to serialization and deserialization issues when crossing language boundaries. Furthermore, the usability of these frameworks has been challenging due to their complex APIs. However, modern big data frameworks backed by C++ offer improved Python APIs with seamless integration with other data analytics systems.

Besides language preference, the widely used big-data systems like Apache Spark have been mainly designed to perform in a lazy execution model where a driver programme takes control of data partitioning and running tasks. This approach bottlenecks the aggregation or reduction tasks, which require all the spawned tasks to communicate with the driver programme to synthesize the final answer. This is a classical problem which can be seen in many big-data systems because of the scheduler semantics.

To address this issue, a few research efforts have been from the big data community to integrate Gang scheduling [23]. Apache Spark has also attempted to incorporate this concept into its schedulers. In addition, research frameworks like Twister [24], Twister2 [25] and Cylon [26] have introduced the usage of MPI for big-data processing by abstracting away the MPI collectives and providing a big-data like APIs for application developers. The performance gains are significant compared to the existing big-data systems.

3.3 Next generation data engineering frameworks

Next-generation data engineering frameworks were built to meet the requirements of modern-day data analytics systems. In Section 2, we learned a few key factors.

- More data leads to good results
- Data analytics systems run on accelerators
- Execution model is BSP or Asynchronous decentralized training (RPC)
- Python is the widely used programming language
- High performance is the key to efficient application development

Big-data systems have been there since 2005, and it has evolved in many ways to meet user requirements. But the major challenge came around 2012 when the data analytics world took over business modeling and solving scientific problems. The rise of neural networks in machine learning made shock waves through the entire data-driven echo system. This is where the aforementioned key factors come into play.

The motto of the last decade was the need to process more significant amounts of data to learn things better. And it began the evolution of modern-day data engineering systems. The key aspects that need an update are;

- Performance
- Usability
- Interoperability with Data Analytics systems
- Low learning curve

Big data systems can process more significant chunks of data but could improve further. The APIs provided in Java, and Scala could be more user-friendly for analysts to use in day-to-day work. There were some efforts to use Python (wrappers on Java/Scala APIs), but they could have been more user-friendly. The schedulers in classical, big-data and modern data analytics systems needed to be aligned, so running end-to-end pipelines was not easy. Also, providing an easier workflow for analysts to perform day-to-day tasks was important.

Figure 3 depicts an approximate performance estimation compared to the usability. Note that this is not a mathematical outcome based on experiments but a collective approximation based on the evolution based on timeline and experience. There is always room for improvement; engineers find ways to improve systems.

Around 2016, a few projects started focusing on better performance and usability. These projects realized the underneath issues in the existing data processing systems. Each project addressed various aspects of the data processing domain, but collectively they can offer a much better solution to data-driven applications.

Figure 4 depicts the main aspects where data engineering frameworks have been evolving. Distributed computing is not a new aspect of data engineering; the older frameworks can do distributed computing. But frameworks like Ray and Cylon have the edge over the existing distributed computing approaches for two main reasons. These two systems have a C++ core backing the performance of sequential operators and communication operations.

Although, before the time of high-performance computing engines, there was an invention called Dask [27] which was mainly designed to provide a set of operators on

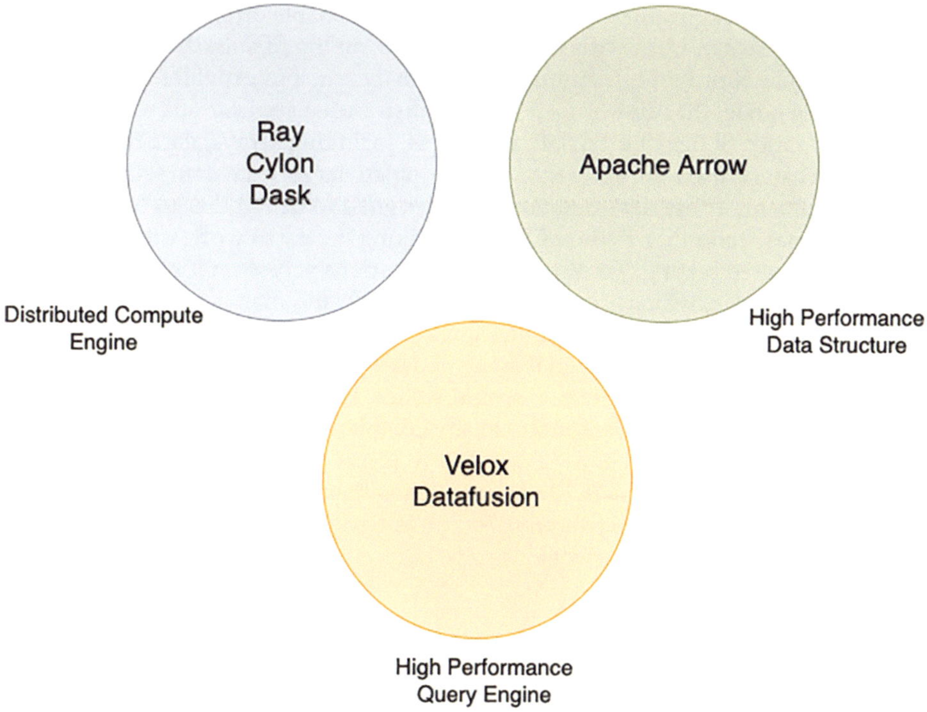

Figure 3.
Approximate estimation on big data systems performance vs. usability.

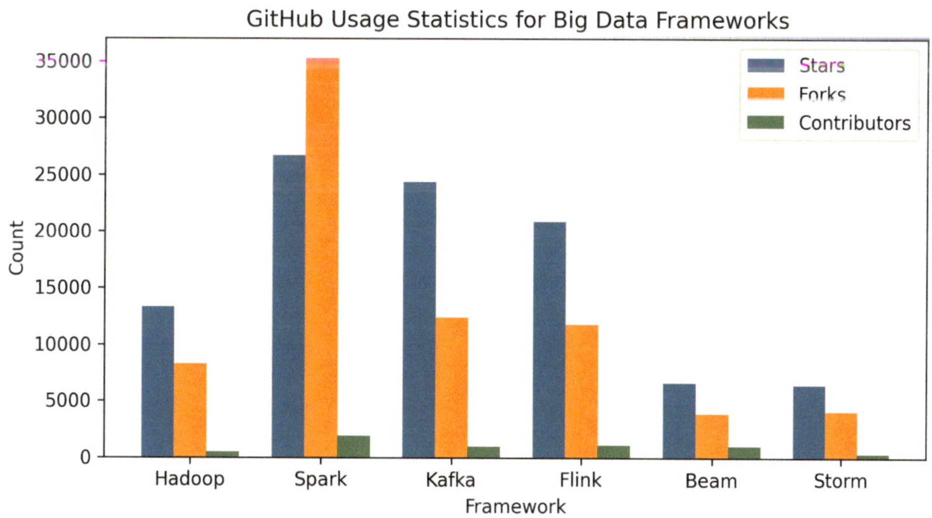

Figure 4.
Modern data engineering.

Pandas [28] to provide distributed computing on a primary workstation. These frameworks were seamlessly integrated with Pandas and Numpy, which made it easier to work with deep learning and machine learning libraries. It later evolved into a distributed framework even to scale in supercomputers. Dask-Distributed is a distributed computing framework that is designed to enable efficient processing of large-scale data sets in Dask with Python. It is built on top of standard Python libraries, such as NumPy [29], Pandas, and Scikit-Learn, and provides a flexible programming model that allows users to write distributed applications with ease. Dask also offers a range of distributed data structures, including arrays, data frames, and bags, which can be used to represent and manipulate large-scale data sets in a distributed environment. These data structures are designed to be familiar to users of NumPy, Pandas, and other Python libraries, making it easy to work with large data sets in a distributed setting. To distribute computations across multiple nodes, Dask uses a task scheduler, which enables users to schedule and manage analyses across a cluster of machines. The scheduler is designed to be fault-tolerant, ensuring that computations continue to run even if some nodes fail or become unavailable. Dask also includes a range of performance optimizations, such as data partitioning and compression, to ensure that computations are completed as efficiently as possible. But one of the critical challenges in the Dask system is it is entirely developed on Python. When running larger workloads, Dask tends to decline performance. The main reasons are the less performance from the Python language and bottlenecks with GIL when running compute-intensive workloads.

```
import dask
large = dask.datasets.timeseries(freq="10s", npartitions=10)
small = dask.datasets.timeseries(freq="1D", dtypes=("z": int))

small = small.repartition(npartitions=1)
result = large.merge(small, how="left", on=["timestamp"])
```

Listing 1.1.
Example Dask Join [30].

A framework called Ray was introduced a few years back to provide an abstraction on distributed training for reinforcement learning and deep learning. Ray provides a flexible programming model that allows developers to write distributed applications in Python with minimal effort. Ray provides an actor based compute model which is easier to scale. It is built on top of the Apache Arrow data format, enabling efficient data transmission without serialization and deserialization between different programming languages or services. It also offers a distributed task scheduler, allowing the users to schedule and manage complex workflows across multiple nodes. Ray is optimized for machine learning applications and provides several built-in libraries and tools for developing and deploying ML models at scale. Overall, Ray aims to make it easy for developers to build and scale distributed applications without worrying about the underlying infrastructure. A commercial version of Ray allows users to work with cloud environments and design applications quite efficiently. An autoscale feature will enable developers to parallelize workloads with trivial command-line arguments. Both Ray and Dask are cloud-friendly tools. But there are systems that were invented way before the cloud was created and are very fast compared to big-data processing systems. These systems are built on high-performance computing (HPC) libraries like MPI [31], OpenMP [32], PGAS [33].

```
import ray
import pandas

ds = ray.data.read_csv("/path-to-iris-data")
ds.show(3)
# Repartition the dataset to 5 blocks.
ds = ds.repartition(5)
# Find rows with sepal.length < 5.5 and petal.length > 3.5.
def transform_batch(df: pandas.DataFrame) -> pandas.DataFrame:
 return df[(df["sepal.length"] < 5.5) (df["petal.length"] > 3.5)]
 # Map processing the dataset.
 ds.map_batches(transform_batch).show()
 # Split the dataset into 2 datasets
 ds.split(2)
 # Sort the dataset by sepal.length.
 ds = ds.sort("sepal.length")
 ds.show(3)
 # Shuffle the dataset.
 ds = ds.random_shuffle()
 ds.show(3)
 # Group by the variety.
 ds.groupby("variety").count().show()
```

Listing 1.2.
Example Ray code for data processing [34].

With an HPC-oriented approach, a framework called Cylon has been specially designed on an MPI-backed collective communication model, which provides high-performance computing capability on Supercomputers. Cylon core communication and compute kernels are written in C++ and extended on Apache Arrow data structures to represent data efficiently. This allows the ability to seamlessly integrate with other Arrow-backed systems and provide efficient data movement from data engineering frameworks to machine learning and deep learning workloads [35–37]. The data structure used underneath is Apache Arrow. It uses Arrow compute kernels to do sequential relational algebra operations while using its partitioned API written on MPI to corresponding distributed operators. Cylon provides APIs in C++, Python and Java. But Cylon has focused on providing more support to Python users by providing a distributed DataFrame library which mimics Pandas but provides distributed operators which abstract away the complex communication algorithms. Cylon supports both CPU and GPU computing. As Apache Arrow for sequential computes, it uses CuDF [38] to do the sequential operators and uses a GPU-supported partition algorithm for the distributed operators. Cylon can be recognized as one of the earliest distributed DataFrame libraries supporting GPU and CPUs. Although Cylon is developed on HPC-based communication models, recently, it has adopted UCX and Gloo as communication backends to enable the workloads to run in cloud-native environments [39].

Big-data frameworks like Apache Spark, Apache Storm, Apache Flink and similar other frameworks have a downside when it comes to being a perfect match for data engineering for data analytics. The gap it left in data engineering is that it is hard to integrate them with the HPC-Pythonic data analytics stack, which runs on HPC hardware with a C++ core backend and an easy-to-use Python API. Cylon fills this gap by providing these high-performance communications and computing APIs to work with data. And enhancing this experience, DataFrames, which Pandas introduced, has become the bedrock of usability. This is one of the unique features of Cylon. Looking into the PyCylon code to do a join in listing 1.3, it is clear that the API is very similar to what Pandas is offering. Pandas has become the go-to tool for data processing for analytic data workloads. But Pandas is a sequential library. Because data analytics frameworks like PyTorch run on BSP-model, they enable MPI-enabled Cylon Dataframes to run seamlessly with data analytics workloads. It is vital when the exploratory analysis is done before designing the production-ready model, where data scientists play with the data to engineer the features and get feedback with the analytical algorithms. It is vital to note that introducing Pythonic HPC solutions is crucial in designing efficient data exploration research to build larger models.

```
from pycylon import read_csv, DataFrame, CylonEnv
from pycylon.net import MPIConfig

config: MPIConfig = MPIConfig()
env: CylonEnv = CylonEnv(config=config, distributed=True)

df1: DataFrame = read_csv('/tmp/csv1.csv')
df2: DataFrame = read_csv('/tmp/csv2.csv')

df3: DataFrame = df1.join(other=df2, on=[0], algorithm="hash", env=env)

print(df3)
```

Listing 1.3.
Example Python code with pycylon.

Both Ray and Cylon provide much better performance compared to Dask. The main reason is the efficiency of computing kernels written on C++ and distributed computing models. Dask also uses a driver-centric distributed computing model, which becomes a bottleneck in running more tasks that need more synchronization than a job with fewer tasks.

Numerous libraries, including Ray and Cylon, utilize Apache Arrow as their underlying data structure. Apache Arrow is a columnar in-memory data format that enables efficient read operations. Its core memory layout is based on a Columnar specification. It allows any framework to adopt the Arrow C Data Interface and extend it to create Arrow-compatible data structures without relying on the entire library. Several libraries have been built upon this columnar specification, including the compute API, dataset API, Flight SQL, Flight RPC, and Acero streaming execution engine. Apache Arrow has gained widespread adoption in various industrial frameworks and academic research, such as Apache Spark, Clickhouse, Dremio, and Polars. Moreover, it supports an extensive range of programming languages, with C++, Python, R, Java, C#, and Go being the most commonly used.

In the heart of data processing, there lie the query engines. Query engines are known to provide a higher-level API for users to run SQL queries or build query plans based on an API. A few key goals of a query engine are fast query execution, scalability, flexibility, concurrency, query optimization, usability, fault tolerance and extensibility. It should have a way to efficiently load, transform, persist and transmit across a wide array of other systems. Modern-day query engines are not built on Java or Scala like in the big-data era; they are built on C++ or Rust for performance requirements. Velox [40] is a high-performing query engine built on C++. It uses Apache Arrow columnar format and has its own set of compute kernels. Velox also supports Presto and Apache Spark query engines. In addition, it also supports feature engineering and data preprocessing in PyTorch. It is a novel system evolving in the high-performance query processing space. Velox is getting more traction in industrial and academic research.

DataFusion [41] is a robust data processing framework that provides users with two distinct methods for creating logical query plans: SQL and DataFrame API. This versatility enables users to choose the most suitable approach for their specific use cases and requirements. The framework also boasts a comprehensive query optimizer that employs advanced optimization techniques to improve the efficiency and performance of query execution. By analyzing and transforming the logical query plans, the optimizer ensures that the most effective execution strategies are used, resulting in faster processing times and better resource utilization. A key feature of DataFusion is its multi-threaded parallel execution engine, designed to process partitioned data sources, such as CSV and Parquet files, with exceptional speed. By using parallelism, the execution engine can distribute the workload across multiple threads or cores, significantly accelerating data processing and analysis. This parallel execution approach is particularly beneficial when working with large datasets, as it can effectively minimize processing times and overcome performance bottlenecks. In summary, DataFusion combines the flexibility of SQL and DataFrame API support, an advanced query optimizer, and a high-performance parallel execution engine to offer a robust and efficient solution for processing and analyzing partitioned data sources like CSV and Parquet files. Its versatile and powerful capabilities make it ideal for various data processing tasks and use cases.

Considering the discussed novel trends and data processing technologies, it is evident that data engineering has evolved from one dimension to another in less than a decade. Researchers and engineers have produced various enhancements to the data engineering stack, and high performance and better usability are the key aspects that have shown progress. Moreover, the number of novel platforms trying toimprove data engineering workloads gives a variety of options for the user. But most of these frameworks have focused on two essential things. Represent and transform data efficiently and seamlessly integrate with the data analytics workloads.

3.4 Anywhere query execution

Among hundreds of frameworks designed for data engineering with uniqueness, each framework is known to be better at specific tasks than the others. In a practical scenario, a few frameworks form data engineering workflows. The main challenge is communicating an idea or simply a query plan so that each framework can understand and do its part. Apache Beam can be recognized as a single framework which unifies frameworks like Apache Spark, Apache Flink, Apache Samza, Google Cloud Dataflow, Twister2, etc. This approach requires writing composite applications by using the Apache Beam API. But in the long run, maintaining such a code base and supporting various platforms is complex and costly. We must run your query anywhere with less overhead and maintenance.

Substrait [42] provides a cross-language specification for data computing operations. In simple terms, once a framework adopts the Substrait specification, it can run a query plan without involving any additional code except for the code required to load a Substrait-based execution plan and execute the framework-native plan. Substrait currently support types, expressions and relations. Under types, data types, type variations and functions (scalar, aggregate, window and Table) are defined. If a particular framework wants to extend upon the existing definitions, such modifications can be done with Substrait. Especially when defining new data types, function signatures and other custom representations of vivid components. These are defined as extensions, which can be defined in a YAML format [43].

At the query execution level, the most exciting component is the Relational algebra support in Substrait. It contains logical and physical relations defined to support most of the widely used relational algebra operators. Read, filter, sort, project, join, set, aggregate, and write are supported, operators. To understand what Substrait can offer, let us evaluate a sample query and how Substrait can represent it. Listing 1.4 shows a SQL query which performs a read operation on a table *LINEITEM* and reading *L_EXTENDEDPRICE*, *L_TAX* and *L_DISCOUNT* columns.

```
SELECT
      t0."L_EXTENDEDPRICE",
      t0."L_TAX",
      t0."L_DISCOUNT"
FROM "LINEITEM" AS t0
```

Listing 1.4.
Example of a Substrait plan in SQL Format.

Substrait does not have a visual format at the moment. It is a protobuf-based non-human readable format which can be sent to Substrait supported framework to execute using the native query engine. But to visualize it with particular readability, we show it using the JSON format. Listing 1.5 offers the Substrait plan in Visual form.

```
{
    "extensionUris": [
        {
            "extensionUriAnchor": 1
        }
    ],
    "relations":  [
        {
            "root": {
                "input": {
                "read": {
                  "common": {
                    "direct": {}
                  },
                  "baseSchema": {
                    "names": [
                      "L_EXTENDEDPRICE",
                      "L_TAX",
                      "L_DISCOUNT"
                  ],
                  "struct": {
                    "types": [
                      {
                          "fp32": {
                      "nullability": "NULLABILITY_NULLABLE"
                          }
                      },
```

```
                        {
                          "fp32": {
                          "nullability": "NULLABILITY_NULLABLE"
                            }
                        },
                        {
                          "fp32": {
                          "nullability": "NULLABILITY_NULLABLE"
                              }
                            }
                        ],
                          "nullability": "NULLABILITY_REQUIRED"
                      }
                    },
                      "namedTable": {
                          "names": [
                              "LINEITEM"
                            ]
                        }
                    }
                },
                "names": [
                      "L_EXTENDEDPRICE",
                      "L_TAX",
                      "L_DISCOUNT"
                ]
            }
        }
    ]
}
```

Listing 1.4.
Example of a Substrait plan in JSON Format.

The Substrait plan contains information about the data source, a *namedTable* representing an in-memory data source. In addition, a file source or a glob can be referred to via URI to define the data source. Since this is a simple read operation, it only shows the *baseSchema* (schema of the data being read, not the schema of the dataset represented in the data source). One of the doubtful questions which arise is why not SQL? SQL is a language used for querying relational data, but it has limitations and lacks sufficient detail for processing. Therefore, modern systems often translate SQL queries into a query plan before executing them. Query plans can have multiple levels and transform, but no standard or open format exists for them. Substrait was created to provide a standard and available form for query plans and works alongside SQL to deliver capabilities that SQL lacks.

Acero (Arrow streaming query engine) [44], Velox, DataFusion and Ibis support Substrait. In terms of support, there are two aspects. There must be a set of producers who can produce Substrait plans and consumers who can execute them. Ibis supports the production and consumption of Substrait plans, while Velox, DataFusion and Acero mainly support Substrait plan consumption. But regarding a producer, the goto tool is *isthmus* [45].

4. Conclusions

The evolution of data engineering over the past decade has been characterized by a trend towards high-performance computing, with frameworks designed to keep up with the evolving demands of the field. The development of diverse frameworks, including Apache Arrow, Apache Parquet, Twister2, Cylon, Velox, and Datafusion, has been essential in providing high-performance data processing on a large scale. The shift towards bulk-synchronous parallel data processing, remote procedure call-based approaches, and extending the usability and extensibility of data analytics workloads have further enhanced the performance of data engineering frameworks. Additionally, the introduction of Substrait has enabled the efficient processing of data across multiple platforms, making it easier for data engineers to build complex data engineering workloads and run queries efficiently. Our study has shown how data analytics has impacted the evolution of data engineering and how modern-day data engineering frameworks have been developed with C++ core compute and communication kernels to facilitate high-performance data processing. Further research can explore the potential of these frameworks in real-world applications and evaluate their performance in handling even larger volumes of data.

Acknowledgements

I am grateful to my wife Kalani, parents, and brother for their unwavering support throughout the writing process. Their guidance and feedback were invaluable. I also thank everyone who contributed to this project.

Abbreviations

SVM	Support Vector Machines
HPC	High-Performance Computing
TPU	Tensor Processing Unit
GPU	Graphics Processing Unit
BSP	Bulk Synchronous Parallel

DOI: http://dx.doi.org/10.5772/intechopen.1001458

Author details

Vibhatha Abeykoon[1*†] and Geoffrey Charles Fox[2†]

1 Indiana University Alumni, Bloomington, IN, USA

2 University of Virginia, Charlottesville, VA, USA

*Address all correspondence to: vibhatha@gmail.com

† These authors contributed equally.

References

[1] Krizhevsky A, Sutskever I, Hinton GE. Imagenet classification with deep convolutional neural networks. Communications of the ACM. 2017; **60**(6):84-90

[2] Introducing ChatGPT. Available from: https://openai.com/blog/chatgpt [Accessed: March 5, 2023]

[3] Hadoop. Apache. Available from: http://hadoop.apache.org [Accessed: November 30, 2022]

[4] Moritz P, Nishihara R, Wang S, Tumanov A, Liaw R, Liang E, et al. Ray: A distributed framework for emerging AI applications. In: 13th USENIX Symposium on Operating Systems Design and Implementation (OSDI 18). 2018. pp. 561-577

[5] Pedregosa F et al. Scikit-learn: Machine learning in python. The Journal of Machine Learning Research. 2011;**12**: 2825-2830

[6] Paszke A, Gross S, Massa F, Lerer A, Bradbury J, Chanan G, et al. Pytorch: An imperative style, high-performance deep learning library. In: Advances in Neural Information Processing Systems. 2019. p. 32

[7] Abadi M, Barham P, Chen J, Chen Z, Davis A, Dean J, et al. Tensorflow: A system for large-scale machine learning. In: Osdi. Vol. 16, No. 2016. 2016. pp. 265-283

[8] Apache MXNet. Amazon Web Services. 2015. Available from: https://mxnet.apache.org/ [Accessed: March 19, 2023]

[9] Jia Y, Shelhamer E, Donahue J, Karayev S, Long J, Girshick R, et al. Caffe: Convolutional architecture for fast feature embedding. In: Proceedings of the 22nd ACM International Conference on Multimedia. 2014. pp. 675-678

[10] Bergstra J, Breuleux O, Bastien F, Lamblin P, Pascanu R, Desjardins G, et al. Theano: A CPU and GPU math expression compiler. In: Proceedings of the Python for Scientific Computing Conference (SciPy). Vol. 4, No. 3. 2010. pp. 1-7

[11] Tokui S, Oono K, Hido S, Clayton J. Chainer: A next-generation open source framework for deep learning. In: Proceedings of Workshop on Machine Learning Systems (LearningSys) in the Twenty-Ninth Annual Conference on Neural Information Processing Systems (NIPS). Vol. 5. 2015. pp. 1-6

[12] TorchScript. PyTorch. 2021. Available from: https://pytorch.org/docs/stable/jit.html [Accessed: March 19, 2023]

[13] Green AI. By Roy Schwartz, Jesse Dodge, Noah A. Smith, Oren Etzioni Communications of the ACM, 2020, Vol. 63 No. 12, Pages 54-63 10.1145/3381831 https://cacm.acm.org/magazines/2020/12/248800-green-ai/fulltext?mobile=false

[14] Mazumder M, Banbury C, Yao X, Karlaš B, Rojas WG, Diamos S, et al. Dataperf: Benchmarks for data-centric ai development. arXiv preprint arXiv: 2207.10062. 2022

[15] Apache Arrow. The Apache Software Foundation. 2016. Available from: https://arrow.apache.org/ [Accessed: March 19, 2023]

[16] Apache Arrow Flight. The Apache Software Foundation. 2020. Available from: https://arrow.apache.org/docs/format/Flight.html [Accessed: March 19, 2023]

[17] Zaharia M et al. Spark: Cluster computing with working sets. HotCloud. 2010;**10**(10–10):95

[18] Apache Kafka. The Apache Software Foundation. 2011. Available from: https://kafka.apache.org/ [Accessed: March 19, 2023]

[19] Apache Flink. The Apache Software Foundation. 2014. Available from: https://flink.apache.org/[Accessed: March 19, 2023]

[20] Apache Beam. The Apache Software Foundation. 2016. Available from: https://beam.apache.org/ [Accessed: March 19, 2023]

[21] Apache Storm. The Apache Software Foundation. 2011. Available from: https://storm.apache.org/ [Accessed: March 19, 2023]

[22] Google Cloud Dataflow. Google. 2014. Available from: https://cloud.google.com/dataflow [Accessed: March 19, 2023]

[23] Feitelson DG, Rudolph L. Gang scheduling performance benefits for fine-grain synchronization. Journal of Parallel and Distributed Computing. 1992;**16**(4):306-318

[24] Ekanayake J, Li H, Zhang B, Gunarathne T, Bae S-H, Qiu J, et al. Twister: A runtime for iterative mapreduce. In: Proceedings of the 19th ACM International Symposium on High Performance Distributed Computing. 2010. pp. 810-818

[25] Kamburugamuve S et al. Twister2: Design of a big data toolkit. Concurrency and Computation: Practice and Experience. 2020;**32**(3):e5189

[26] Widanage C, Perera N, Abeykoon V, Kamburugamuve S, Kanewala TA, Maithree H, et al. High performance data engineering everywhere. In: 2020 IEEE International Conference on Smart Data Services (SMDS), Remote. IEEE; 19 Oct 2020. pp. 122-132

[27] Rocklin M. Dask: Parallel computation with blocked algorithms and task scheduling. In: Proceedings of the 14th Python in Science Conference. Vol. 130. Austin, TX: SciPy; 2015

[28] McKinney W. Pandas: A foundational python library for data analysis and statistics. Python for high performance and scientific computing. 2011;**14**(9):1-9

[29] Van Der Walt S, Colbert SC, Varoquaux G. The NumPy array: A structure for efficient numerical computation. Computing in Science & Engineering. 2011;**13**(2):22-30

[30] Dask Join. Available from: https://docs.dask.org/en/stable/generated/dask.dataframe.DataFrame.join.html

[31] Dongarra JJ, Otto SW, Snir M, Walker D, et al. An introduction to the MPI standard. Communications of the ACM. 1995;**18**

[32] Dagum L, Menon R. OpenMP: An industry standard API for shared-memory programming. IEEE Computational Science and Engineering. 1998;**5**(1):46-55

[33] Chapman B, Curtis T, Pophale S, Poole S, Kuehn J, Koelbel C, et al. Introducing OpenSHMEM: SHMEM for the PGAS community. In: Proceedings of the Fourth Conference on Partitioned Global Address Space Programming Model. 2010. pp. 1-3

[34] Ray Dataset. Available from: https://docs.ray.io/en/latest/data/transforming-datasets.html

[35] Kamburugamuve S, Widanage C, Perera N, Abeykoon V, Uyar A, Kanewala TA, et al. Hptmt: Operator-based architecture for scalable high-performance data-intensive frameworks. In: 2021 IEEE 14th International Conference on Cloud Computing (CLOUD), Chicago, IL, USA. IEEE; 2021. pp. 228-239

[36] Abeykoon V, Perera N, Widanage C, Kamburugamuve S, Kanewala TA, Maithree H, et al. Data engineering for hpc with python. In: 2020 IEEE/ACM 9th Workshop on Python for High-Performance and Scientific Computing (PyHPC), GA, USA. IEEE; 2020. pp. 13-21

[37] Perera N, Abeykoon V, Widanage C, Kamburugamuve S, Kanewala TA, Wickramasinghe P, et al. A fast, scalable, universal approach for distributed data aggregations. In: 2020 IEEE International Conference on Big Data (Big Data), GA, USA. IEEE; 2020. pp. 2691-2698

[38] cuDF. NVIDIA Corporation. 2021. Available from: https://github.com/rapidsai/cudf

[39] Perera N, Kamburugamuve S, Widanage C, Abeykoon V, Uyar A, Shan K, et al. High performance dataframes from parallel processing patterns. arXiv preprint arXiv: 2209.06146. 2022

[40] Pedreira P et al. Velox: meta's unified execution engine. Proceedings of the VLDB Endowment. 2022;**15**(12): 3372-3384

[41] Datafusion. Apache Arrow. Available from: https://arrow.apache.org/datafusion/ [Retrieved: February 28, 2023]

[42] Substrait. Available from: https://substrait.io/

[43] Substrait YAML Spec. Available from: https://github.com/substrait-io/substrait/blob/main/text/simple_extensions_schema.yaml

[44] Acero, Streaming Execution Engine for Apache Arrow. Available from: https://arrow.apache.org/docs/cpp/streaming_execution.html

[45] Substrait Isthmus. Available from: https://github.com/substrait-io/substrait-java/tree/main/isthmus

Section 2

Case Studies

Chapter 4

GeoImage Workflow Editing Resources: GIWER

Istvan Elek

Abstract

While various open-source and commercial image-processing software exist, and they have high-level image-processing skills, there is no flexibility to assemble custom workflows where the experience of the evaluator can be built into the system. Therefore we have developed the Giwer (acronym for GeoImage Workflow Editing Resources, pronounce gaiver) open-source software package for handling, processing and analyzing drone images with the possibility of flexible workflow creation. Giwer has many image-processing functions that can handle traditional RGB, greyscale and hyperspectral images and apply them for their interpretation for any number of images. Giwer has further useful capabilities to organize images and their metadata (EXIF) into a database to navigate thousands of images. Users can compile their own workflows based on the available image-processing functions. Thus the created workflow involves users' knowledge and experiences. The workflows process any number of images at once. This approach can accelerate the performance if you have a working algorithm group.

Keywords: remote sensing, drone image processing, open-source, workflow editing, hyperspectral images, GIS

1. Introduction

Nowadays, drone technology is already present in almost all areas of life. This is especially true for precision agriculture, environmental protection, nature conservation, the police and the military. There are world-famous programmes in this field that are capable of processing drone images. Among the commercial software, ENVI [1] and ArcGIS [2] stand out, but there are also notable products among the open-source software packages, such as SAGA [3] or QGIS [4]. Although the mentioned programmes have outstanding functions, creating your own processing procedures with them is difficult or impossible.

That is why we decided to create a programme system that supports the creation of almost any processing process from among the functions available. Ultimately, we can create arbitrary workflows [5, 6] incorporating the own knowledge and experience of the image interpretation expert into the process. Since we have complete control over the source code, any procedures can be implemented. It is the Giwer [7] system.

Furthermore, we developed to run these workflows for an arbitrary number of images, so that a procedure worked out for an area can be applied to all available images. This can significantly reduce the processing time of images of the area.

2. The Giwer system

Our objective was to develop a programme package that can process mainly drone images, but of course, it can also handle satellite images. The other goal was to serve workflow editing capabilities, where users can create their own workflows. This workflow can be applied to a single image and to any number of images. This is the project concept where you can create an image group including any number of images.

The further aim was to implement an image catalog (its name is Catalogue) where the images are stored in a database with many metadata of the images. If you have hundreds of images, it is not easy to navigate among images and find the right one if you are not able to create useful queries.

Giwer uses its own data structure to be the fastest performance. All original image formats (tif, geotif, jpg, bil, ENVI bil and cub) should be converted to Giwer format (gwr) before using the whole functionality of Giwer (**Figure 1**).

The heart of Giwer is a framework which organizes the different subsystems. You can run any subsystem in parallel. Let us see the subsystems.

2.1 Catalog

It organizes a large number of images into a database. The Catalog stores raw images in a database (Sqlite). Catalog reads and stores many images and image parameters from their exif data coming directly from the media of drone. Additionally, it also provides storage options in some interactive fields (**Figure 2**).

Some camera, such as Micasense, produces as many image files as many bands as it has. Unfortunately, there is a little shift between the different images. It has to be corrected for the proper interpretation (**Figure 3**). Only after this correction should image files be converted to gwr format.

2.1.1 Experiences with MicaSense camera

Multispectral cameras with multiple lenses contain aligning mistakes in the images for the following reasons [8, 9]:

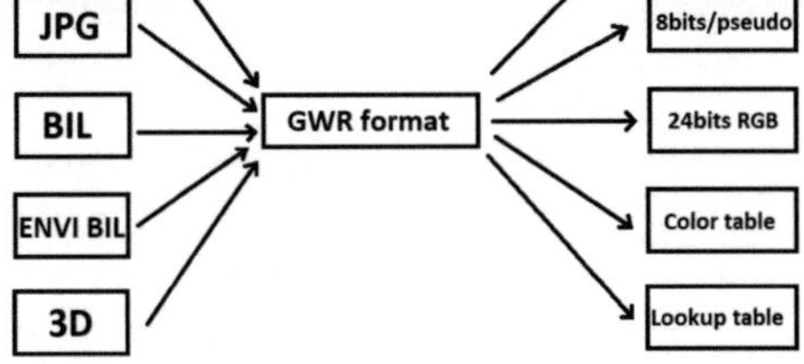

Figure 1.
Source data should be converted to format gwr, which is a byte array from band to band.

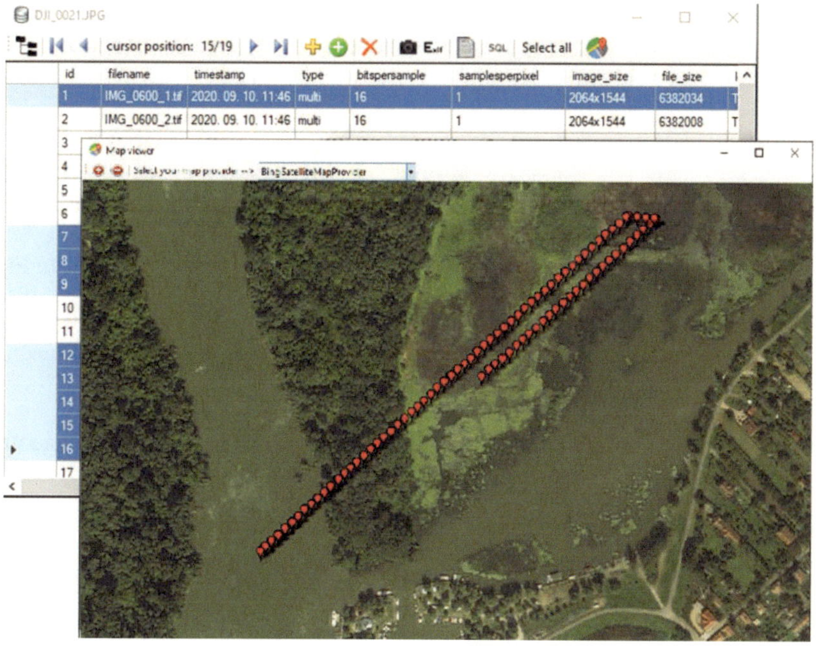

Figure 2.
A screenshot of the Catalog can be seen in this figure. Tabular data are in the background, and the flight trajectory is in the foreground.

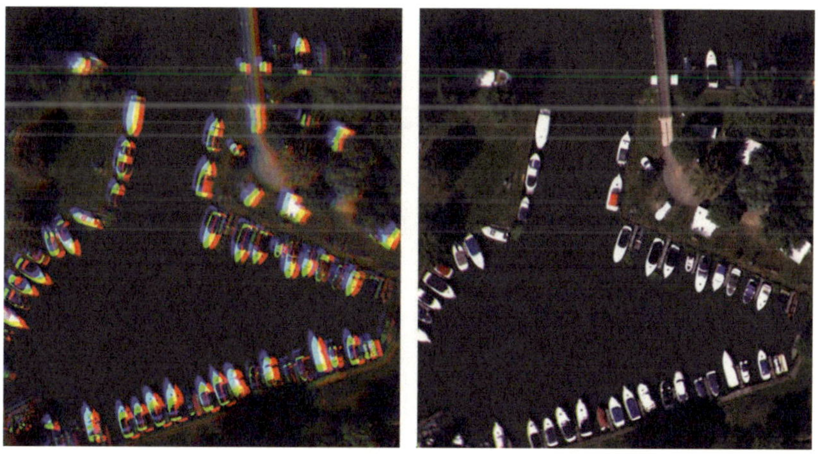

Figure 3.
An RGB image before (left side) and after (right side) alignment correction. The alignment correction is based on the affine transformation since we must eliminate shifts and rotations.

- The Micasense cameras produce as many files as many bands as they have. We have installed RGB and multispectral cameras on the drone. There was no problem with RGB, but the images of the multispectral camera have some discrepancies.

- Based on the construction of the multispectral camera, it has as many bands as many lenses. Thus the image files have a little shift related to each other, making the interpretation of images impossible. In addition, the individual camera positions have very little rotation, and lenses have different distortion characteristics, which also should be corrected [9–11].

- Unfortunately, the built-in camera-GPSs are not accurate enough, i.e. the central image coordinates are proper for outlining the flight path but not for making mosaic from images. We attempted to resolve this problem with cross-correlation between images, but the result was not accomplished. Therefore we applied affine transformation to resolve this problem, and the result was excellent [7]. In **Figure 8**, we can see the original image and the result after affine transformation.

2.2 DataStock

It is an interactive image-processing system. We have implemented large number of image-processing functions that can be accessed via the menu system. Its main functions are the following

- Loads images in different formats: gwr, bil, ENVI Bil, tif, geotif, jpg, cub.
- Loads 8, 16, 24 and 48 bits images, from 3 banded RGB to 250 banded hyperspectral images.
- Creates an RGB image from any of three tracks (**Figure 4**).
- Histogram equalization and drawing (**Figure 4**).

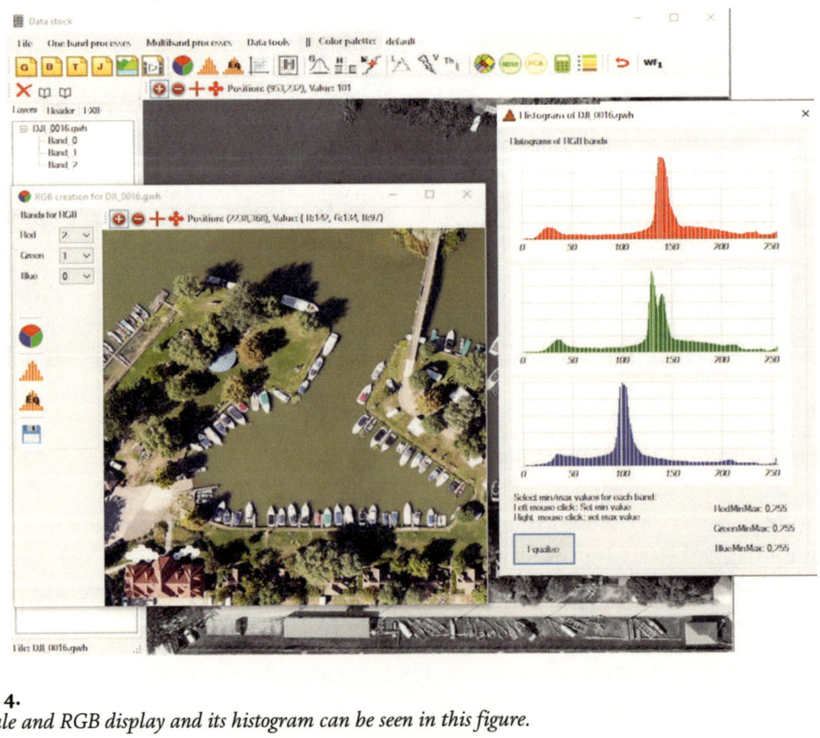

Figure 4.
Greyscale and RGB display and its histogram can be seen in this figure.

- Cross-plot drawing from any of two bands, such as red and green (**Figure 5**).
- Handling file header (display/edit).
- Apply available functions to process images.
- NDVI and PCA calculation (**Figures 6** and 7).

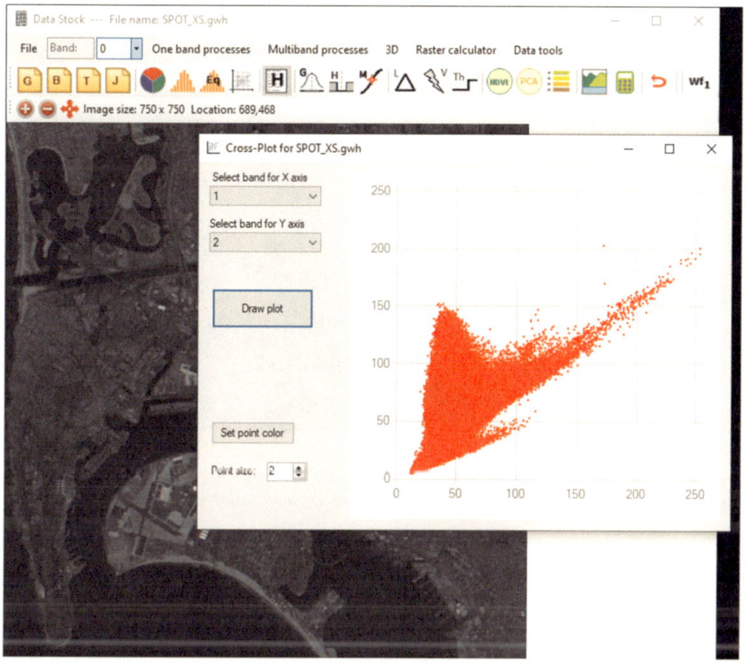

Figure 5.
A cross-plot display where the X axis is the intensity of band#1, and the Y axis is the intensity of band#2.

Figure 6.
This is the result of the NDVI calculation. On the left side, an RGB image can be seen, and on the right side, the NDVI.

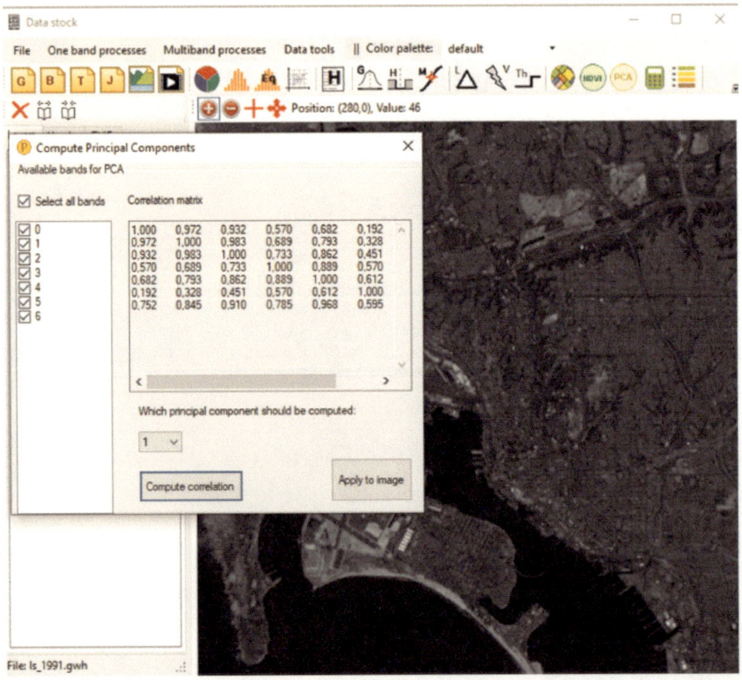

Figure 7.
This is the result of the first principal component computation. You can also see the correlation matrix of images.

- Display 3D data with greyscale, hypsometric or user-defined lookup table (**Figure 8**).

- Raster calculator: querying according to arbitrary, user-defined conditions, and a special graphic selection based on cross-plot technique (**Figure 9**).

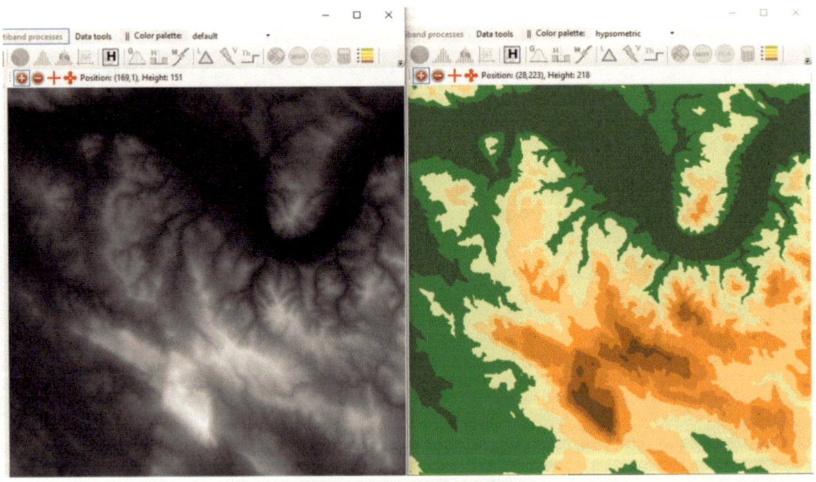

Figure 8.
A greyscale image (left) and a hypsometric version of it (right) can be seen here. The location is the Danube bend.

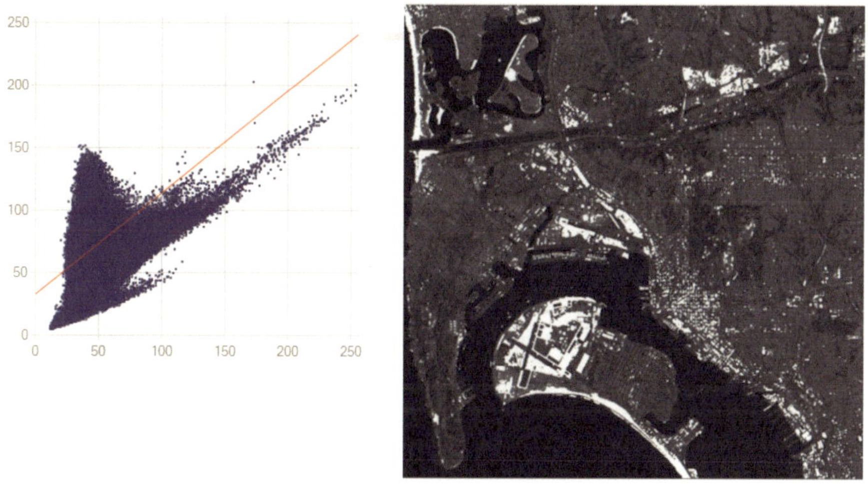

Figure 9.
Graphic selection with intensity cross-plot. If you select the pixels above the red lines, you can see them in white color on the right side.

- The combination of images (add, average, exclusive, subtract, etc.) can be seen in **Figure 10**. Any image-processing result can be combined with other images or results if they are overlapped. Disjunct images cannot be combined.

- Conversion between data formats

- Filter bank (e.g. smoothing, edge detection, median filtering, etc.). **Figure 11** is an illustration of edge detection

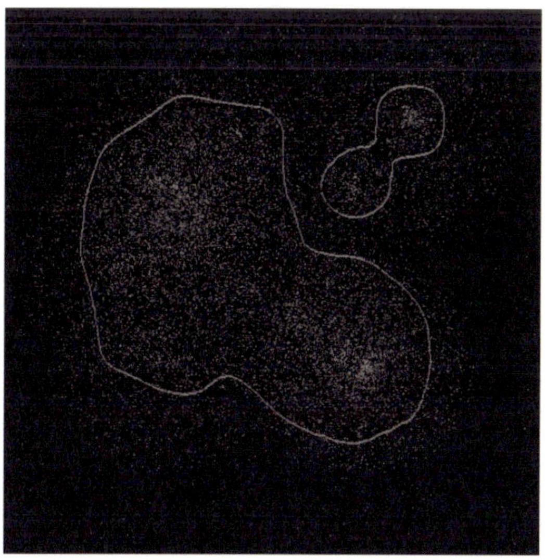

Figure 10.
This is a composite image where the point clouds and their automatically detected boundaries were combined.

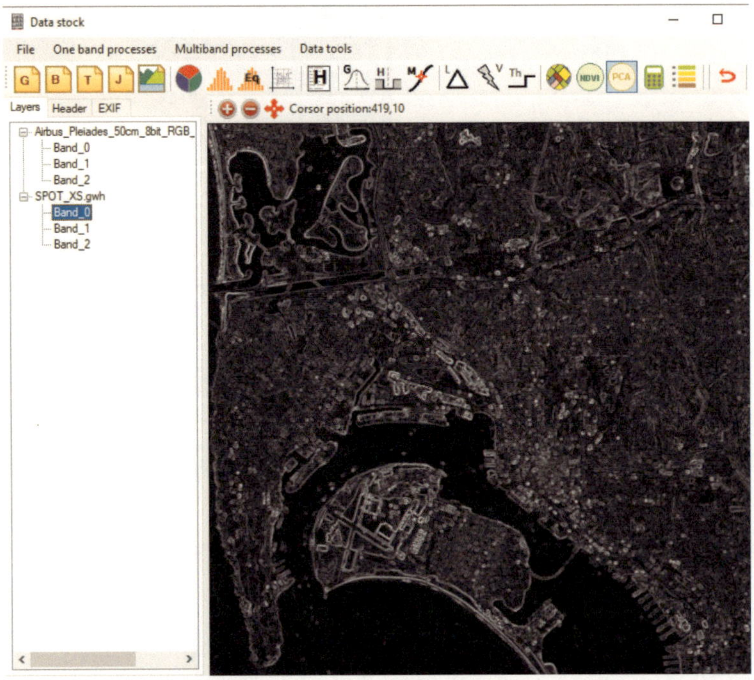

Figure 11.
A result of edge detection on a satellite image.

- Classification (**Figure 12**). Many clustering methods are included in Giwer, such as K-means, K-means (Multi threaded), K-means (Manual), and Random forest.

- Editing spectrum banks (**Figure 13**). The local and global spectrum banks support the interpretation of images, especially in agriculture [12].

- Classification by spectrums

2.3 WorkflowBuilder

This is a workflow editor. From the available functions, the arbitrary workflow can be compiled, so the user can create their own processing procedure based on their individual knowledge, experience and creativity (**Figure 14**). Sometimes it can be useful if we combine images with Lidar data. This part of Giwer has not been completed, but it will in the future [13–15].

3. Summary, future

The implementation of Giwer proved that this software package for organizing, categorizing and batch processing drone images is useful for university labs and projects for different experimental work in the field such agriculture, nature conservation

DOI: http://dx.doi.org/10.5772/intechopen.1001297

Figure 12.
A result of isodata clustering.

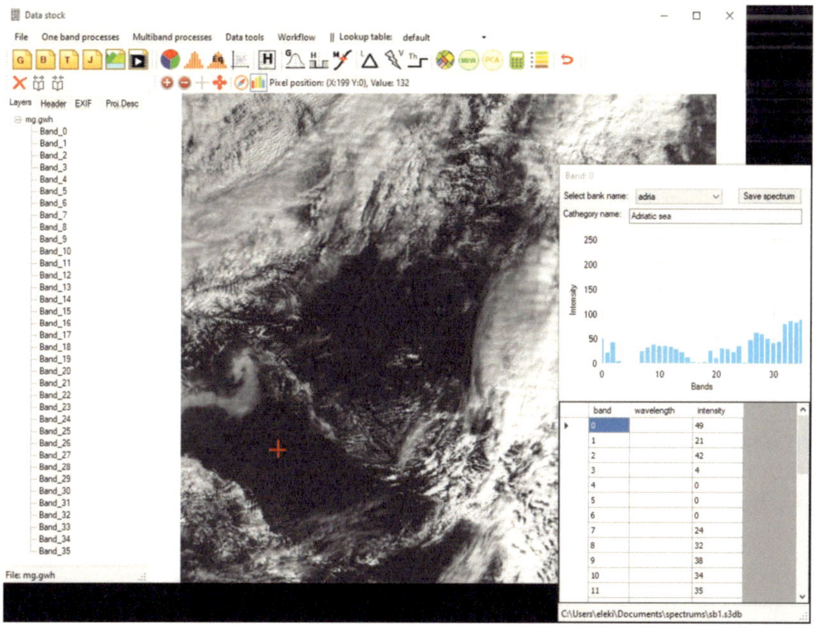

Figure 13.
An image and its spectrum in the selected point (red cross) can be seen in this figure if you select a pixel with the mouse click. The spectrum immediate quick look is also possible if you are moving the cursor on the image.

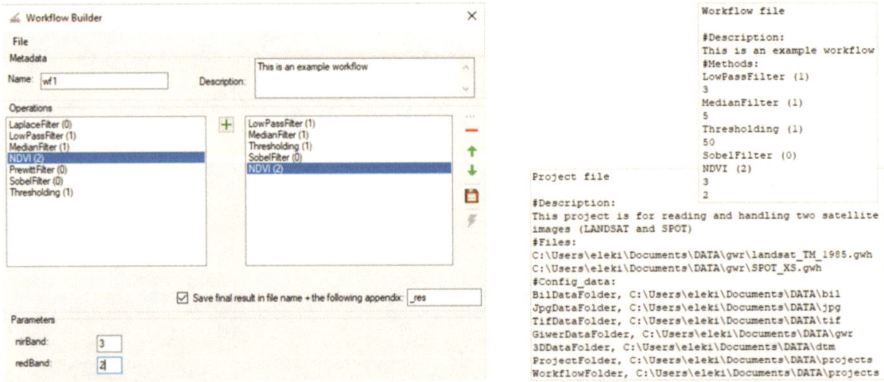

Figure 14.
This figure illustrates the WorkflowBuilder. On gray background, you can see a subset of the available functions (left list) and the selected functions to put them into the new workflow (right list). Further, you can see the edited workflow as a text file with function names and the required parameters (upper right part). On the lower right part of this figure, the project file can be seen, which contains the image files which are the target of the workflow.

and environment protection. Giwer is also a good tool to analyze images from satellites and planes.

The recent stable version of Giwer is 1.1. We continue the development focused on the hyperspectral and making mosaics. In addition, we plan to add further cameras with special preprocessing functions.

4. Other information

Giwer was written in C# at ELTE Eotvos Lorand University, Faculty of Informatics. Every subsystem, such as DataStock, Catalog, WorkflowBuilder runs alone in parallel as well without a frame programme.

Giwer is an open-source programme package with GN3 license. It can be found in GitHub, where you can download the source code (https://github.com/istvan-elek/Giwer). Executables are downloadable from http://mapw.elte.hu/giwer. A detailed user's guide is available for help to install and use the system.

Acknowledgements

Project no. ED_18-1-2019-0030 (application-specific, highly reliable IT solutions) has been implemented with support from the National Research, Development and Innovation Fund of Hungary, financed under the Thematic Excellence Programme funding scheme.

Author details

Istvan Elek
ELTE Eotvos Lorand University Faculty of Informatics, Budapest, Hungary

*Address all correspondence to: elek@inf.elte.hu

References

[1] Image Analysis with ENVI. http://gmtgis.com/products/image-analysis-envi/

[2] ArcGIS Documentation. https://desktop.arcgis.com/en/arcmap/

[3] System for Automated Geoscientific Analyses. http://www.saga-gis.org/

[4] QGIS Documentation. https://docs.qgis.org/3.16/en/docs/

[5] Microsoft Dynamic Programming. https://learn.microsoft.com/en-us/dotnet/framework/reflection-and-codedom/emitting-dynamic-methods-and-assemblies

[6] Microsoft Documentation on Reflection. https://learn.microsoft.com/en-us/dotnet/framework/reflection-and-codedom/reflection

[7] Elek I, Cserep M. Processing drone images with the open source Giwer software package, lecture notes in networks and systems 359. In: Proceedings of the Future Technologies Conference (FTC). Vancouver; 2021. Available from: https://link.springer.com/chapter/10.1007/978-3-030-89880-9_15

[8] Bay H, Ess A, Tuytelaars T, Van Gool L. Speeded-up robust features (SURF). Computer Vision and Image Understanding. 2008;**110**:346-359

[9] Jhan J-P, Rau J-Y, Haala N. Robust and adaptive band-to-band image transform of UAS miniature multi-lens multispectral camera. ISPRS Journal of Photogrammetry and Remote Sensing. 2018;**137**:47-60

[10] Jhan JP, Rau JY. A normalized surf for multispectral image matching and band co-registration. In: The International Archives of the Photogrammetry, Remote Sensing and Spatial Information Sciences. Vol. XLII-2/W13. Enschede, The Netherlands: ISPRS Geospatial Week 2019; 2019

[11] Ye Y, Shan J. A local descriptor based registration method for multispectral remote sensing images with non-linear intensity differences. ISPRS Journal of Photogrammetry and Remote Sensing. 2014;**90**:83-95

[12] Alexy M, Jung A, Molnar B. Information technology drivers in smart farming management systems. In: Subhan D, Hakoomat A, Rahul D, editors. Smart Farming [Working Title]. London, United Kingdom/England: IntechOpen; 2022

[13] Cserep M, Lindenbergh R. Distributed processing of Dutch AHN laser altimetry changes of the built-up area. International Journal of Applied Earth Observation and Geoinformation. 2023;**116**, art. 103174:12

[14] Cserep M, Demjan A, Mayer F, Tabori B, Hudoba P. Effective railroad fragmentation and infrastructure recognition based on dense LIDAR point clouds. ISPRS annals of the photogrammetry. Remote Sensing and Spatial Information Sciences. 2022;**2**: 103-109

[15] Fekete A, Cserep M. Tree segmentation and change detection of large urban areas based on airborne LiDAR. Computers & Geosciences. 2021; **156**:art. 104900

Chapter 5

Perspective Chapter: A New Bivariate Inverted Nakagami Distribution – Properties and Applications

Aliyu Ismail Ishaq, Abubakar Usman, Ahmad Abubakar Suleiman, Mahmod Othman, Hanita Daud, Rajalingam Sokkalingam, Uthumporn Panitanarak and Muhammad Azrin Ahmad

Abstract

In this work, a new bivariate inverted Nakagami distribution that can be used to model real-world datasets has been investigated. The newly developed bivariate distribution's cumulative distribution function and probability density function are defined. The bivariate distribution derives from the Farlie Gumbel Morgenstern, and the marginal density functions are also determined. Some fundamental estimation techniques, such as maximum-likelihood estimation and inference functions for margins, are used to derive the parameters of its estimates. Applications to real-world datasets pertaining to kidney infection diseases and the UEFA Champions' League group stage for the seasons 2004–2005 and 2005–2006 help to assess the efficacy of the proposed distribution.

Keywords: bivariate inverted Nakagami, Farlie Gumbel Morgenstern, inverted Nakagami, marginal density functions, maximum likelihood estimation

1. Introduction

Over the past decades, many researchers have attempted to introduce new of probability distributions that provide better flexibility than the traditional ones. However, several of these distributions are inappropriate for modeling different characteristics of real data. Therefore, there is a need to develop more flexible distributions, particularly in practical domains including finance, environment, health, and engineering. This study proposed a novel multivariate probability distribution known as the bivariate inverted Nakagami distribution. This bivariate distribution was introduced from the inverse Nakagami distribution. The proposed distribution can serve as alternative to various current distributions, such as the traditional Nakagami and inverse Nakagami distributions and many others.

2. Background

The problem of developing novel families of continuous bivariate distributions is one of the significant and current research topics in probability and statistics. This is due to the limitations of the existing distributions that capture the true behavior of many real phenomena found in a broad variety of practical domains. The Nakagami distribution was introduced recently [1]. This distribution has been applied in ultrasound images [2], microwave hyperthermia [3], cataract stiffness [4], and many other applications. The Nakagami distribution is a probability distribution with two parameters that is related to the gamma distribution. This distribution can be used quite effectively in modeling many empirical datasets [5], especially in communications engineering and mobile radio [6–9]. Nakagami distribution has also found important applications in wind speed [10], medical sciences [11, 12], and hydrologic engineering [13–15]. Other important applications of this distribution are in medical image processing [16, 17], seismological analysis [18], and engineering [14]. This distribution has been used to model the hazard rate in reliability theories because of its memory less property. It has been shown that the Nakagami distribution is a more appropriate function to evaluate the reliability of electrical components compared to the Weibull and Gamma distributions [19].

Due to the successful use of Nakagami distribution in different fields, several researchers have explored the applicability of this distribution. For example, the Nakagami distribution was used [20] to evaluate the ablated region induced by focused ultrasound exposures at different acoustic power levels in transparent tissue-mimicking phantoms. Schwartz et al. [21] developed analytic and bootstrap bias-corrected maximum-likelihood estimators for the shape parameter of the Nakagami distribution. The relationship between the Nakagami distribution and other distributions such as the gamma distribution, the Rayleigh distribution, the Weibull distribution, the chi-square distribution, and the exponential distribution was explored [22]. The study suggested that through the gamma distribution, it is much easier to derive the moments of a Nakagami random variable. The Bayesian estimators of the scale parameter of the Nakagami distribution were derived [23]. The performance of the estimator was evaluated based on the relative posterior risk. The maximum-likelihood estimates for the Nakagami distribution have been compared with other estimators [24]. Recently, the Bayesian method of estimation is used in order to estimate the scale parameter of the Nakagami distribution by using Jeffreys', Extension of Jeffreys', and Quasi priors under three different loss functions [24]. Some of the distributional properties and reliability characteristics of this distribution are discussed [25].
The length-biased form of the Nakagami distribution was introduced by [26]. The new length-biased Nakagami distribution was applied [27] to generate a survival model.

The inverse Nakagami (inverted Nakagami) distribution is proposed [28]. This distribution is the reciprocal of the Nakagami model that plays an important role in the general areas of medical, communication engineering, hydrological sciences, and reliability systems. The proposed model is useful to describe devices that are subjected to high stress, providing a high failure rate after a short repair time.

In many practical problems, multivariate lifetime data arise frequently, and in these situations, it is important to consider different multivariate models that could be used to model such multivariate lifetime data. Several authors, for example, [29–33], have considered the problem of proposing general multivariate models with given marginal distributions. There are very few multivariate distributions in the recent statistical literature. These include the bivariate Kumaraswamy distribution [34], the bivariate Poisson exponential-exponential distribution [35], and the bivariate alpha power exponential distribution [36]. For a bivariate model having given marginals to be useful in practical situations, it is important and desirable that the model can be handled with mathematical ease and that any parameter(s) incorporated in the model lends itself to some important physical representation, for example, the measure of location or scale or an association between components, etc. [37].

A random variable T is said to follow an inverted Nakagami distribution with shape parameters a and b if it's cumulative distribution function (cdf) and probability density function (pdf) are respectively given as

$$F(t;a,b) = 1 - \frac{\gamma\left(a, \frac{a}{bt^2}\right)}{\Gamma(a)}, \qquad a,b;t > 0 \tag{1}$$

and

$$f(t;a,b) = \frac{2}{\Gamma(a)}\left(\frac{a}{b}\right)^a t^{-2a-1}\exp\left(-\frac{a}{bt^2}\right), \qquad a,b;t > 0 \tag{2}$$

A bivariate Nakagami distribution with identical fading parameters was first presented in [1]. In Ref. [38], this restriction was raised and a bivariate Nakagami distribution with arbitrary fading parameters was derived. Recently, a bivariate Nakagami distribution with arbitrary correlation and fading parameters was studied [39]. The primary reason for this expansion was to derive the joint moment generating function, joint probability density function, joint cumulative distribution function, power correlation coefficient, and several statistics related to the signal-to-noise ratio at the output of the selection combiner, namely, outage probability, probability density function, mean, and among other expressions. A new multivariate Nakagami distribution with arbitrary correlation and fading parameters was introduced [40] to obtain the joint probability density function for the Nakagami distribution generated from correlated Gaussian random variables based on an arbitrary correlation matrix and different fading parameters.

Recently, many researchers considered the bivariate extension of the probability distributions, such as Yang et al. [41] presented a class of multivariate copulas whose two-dimensional marginals belong to the family of bivariate Fréchet copulas. Myrhaug and Leira [42] discussed the bivariate Fréchet distribution, which is obtained by transforming a bivariate Rayleigh distribution. Zheng et al. [43] discussed the bivariate Fréchet copula as a mixture of three simple structures co-monotonicity, independence, and counter-monotonicity. A copula is a convenient approach to describe a multivariate distribution with a dependence structure. Nelsen [44] introduced copulas as following; copula is a function that joins multivariate distribution functions with uniform [0, 1] margins. Sklar [45] introduced the pdf and cdf for the two dimension copula as follows, consider the two random variables T_1 and T_2, with distribution functions $F_1(t_1)$ and $F_2(t_2)$ respectively, then the cdf and pdf for bivariate copula are respectively given as

$$F(t_1,t_2) = C(F_1(t_1),F_2(t_2)), \tag{3}$$

$$f(t_1,t_2) = f_1(t_1)f_2(t_2)c(F_1(t_1),F_2(t_2)). \tag{4}$$

where $C(F_1(t_1), F_2(t_2))$ and $c(F_1(t_1), F_2(t_2))$ represents the copulas function for the cdf and pdf of the Bivariate function.

In this case, $F(t_i)$ and $f(t_i)$ for $i = 1, 2$ represents the cdf and pdf of the Inverted Nakagami distribution.

Many copulas had been defined based on Eqs. (3) and (4) such as Farlie-Gumbel-Morgenstern (FGM), Ali-MikhailHaq (AMH), and Plackett. The FGM copula is one of the most popular parametric families of copulas, the family was first introduced [46]. Almetwally et al. [47] used the FGM copula to introduce the bivariate Weibull distribution. Ali et al. [37] proposed an AMH copula, and Kumar [48] discussed the correlation coefficient of the AMH copula by Spearman and Kendall. Almetwally and Muhammed [49] studied the bivariate extension of the Fréchet distribution based on FGM and AMH copula functions and discussed their statistical properties. The Plackett copula was introduced [50] to construct a class of bivariate distributions from given margins. This class contains the known boundary distributions and the members corresponding to independent random variables.

The need for an accurate and effective estimating method for real life data using probability distribution is of great importance. This chapter presents a novel bivariate inverted Nakagami, which provides greater accuracy and flexibility in fitting real life data in a broad variety of practical domains.

In this chapter, we examine and describe the statistical characteristics of the novel bivariate inverted Nakagami distribution based on the FGM copula function. Different estimation techniques are used to estimate the parameters for the bivariate-inverted Nakagami distribution.

2.1 Motivation of the chapter

The bivariate probability distributions are set of distributions proposed to foster new hybridized probability distributions with the intent of expanding the modeling capacity of classical probability distributions. This work attempts to improve the classical Nakagami and inverse Nakagami of distributions for modelimg real life data.

2.2 Challenges of the topic

The Nakagami and inverse distributions seem to be flexible but has not been fully explored in statistical literature and several of their properties have not been studied.

2.3 Significance/implication

This research work developed a bivariate distribution capable of handling skewness and leptokurtic behavior in most datasets in different fields such as medicine, engineering, finance and economics. It also shown that noticeable improvements are made when the bivariate inverted Nakagami distribution is used and tested among the traditional Nakagami models.

The remaining section of this chapter is structured as follows: In Section 2, a bivariate-inverted Nakagami distribution has been identified. Section 3 discusses parameter estimation techniques for the bivariate inverted Nakagami distributions. Applications to two real-world datasets are provided in Section 4, and Section 5 addresses the conclusion of a few remarks for the bivariate-inverted Nakagami model.

3. Bivariate-inverted Nakagami distribution

Bivariate-inverted Nakagami (BIN) distribution can be obtained based on copula function by considering the cdf and pdf defined respectively in Eqs. (3) and (4) presented as

$$F(t_1,t_2) = C\left(1-\gamma_1\left(a_1, \frac{a_1}{b_1{t_1}^2}\right),\ 1-\gamma_2\left(a_2, \frac{a_2}{b_2{t_2}^2}\right)\right) \tag{5}$$

which is the cdf of the BIN distribution, where $\gamma_i\left(a_i, \frac{a_i}{b_i{t_i}^2}\right) = \frac{\gamma\left(a_i, \frac{a_i}{b_i{t_i}^2}\right)}{\Gamma(a_i)}$ for $i = 1, 2$. The pdf corresponding to Eq. (5) is obtained as

$$f(t_1,t_2) = \frac{4}{\Gamma(a_1)\Gamma(a_2)}\left(\frac{a_1}{b_1}\right)^{a_1}\left(\frac{a_2}{b_2}\right)^{a_2} {t_1}^{-2a_1-1}{t_2}^{-2a_2-1}\exp\left(-\frac{a_1}{b_1{t_1}^2}\right)\exp\left(-\frac{a_2}{b_2{t_2}^2}\right)c\left(1-\gamma_1\left(a_1, \frac{a_1}{b_1{t_1}^2}\right), 1-\gamma_2\left(a_2, \frac{a_2}{b_2{t_2}^2}\right)\right) \tag{6}$$

According to Refs. [45–47], the cdf and pdf presented in Eqs. (3) and (4) can be defined as

$$C(\theta,\lambda) = \theta\lambda\{1+\alpha(1-\theta)(1-\lambda)\} \tag{7}$$

and

$$c(\theta,\lambda) = \{1+\alpha(1-2\theta)(1-2\lambda)\}, \tag{8}$$

which is the FGM copula class, where $\theta = F_1(t_1)$, $\lambda = F_2(t_2)$; then $v, w \in \mathrm{I}$ for $\mathrm{I} = [0,1]$ and $\alpha \in [-1,1]$, and this serves as the dependence parameter, likewise an independence parameter if $\alpha = 0$.

As defined in Eqs. (7) and (8), the cdf and pdf of the new bivariate-inverted Nakagami distribution can be obtained from Eqs. (5) and (7) as

$$F(t_1,t_2) = \left\{1-\gamma_1\left(a_1, \frac{a_1}{b_1{t_1}^2}\right)\right\}\left\{1-\gamma_2\left(a_2, \frac{a_2}{b_2{t_2}^2}\right)\right\}\left\{1+\alpha\left(\gamma_1\left(a_1, \frac{a_1}{b_1{t_1}^2}\right)\right)\left(\gamma_2\left(a_2, \frac{a_2}{b_2{t_2}^2}\right)\right)\right\} \tag{9}$$

and

$$\begin{aligned} f(t_1,t_2) = {} & \frac{4}{\Gamma(a_1)\Gamma(a_2)}\left(\frac{a_1}{b_1}\right)^{a_1}\left(\frac{a_2}{b_2}\right)^{a_2} {t_1}^{-2a_1-1}{t_2}^{-2a_2-1}\exp\left(-\frac{a_1}{b_1{t_1}^2}\right)\exp\left(-\frac{a_2}{b_2{t_2}^2}\right) \\ & \times\left\{1+\alpha\left(1-2\left(1-\gamma_1\left(a_1, \frac{a_1}{b_1{t_1}^2}\right)\right)\right)\left(1-2\left(1-\gamma_2\left(a_2, \frac{a_2}{b_2{t_2}^2}\right)\right)\right)\right\} \end{aligned} \tag{10}$$

4. Parameter estimation of the copula-based model

In this section, the maximum-likelihood estimation (MLE) and Inference functions for margins (IMF) are employed in estimating the parameters of the bivariate-inverted Nakagami distribution.

4.1 Estimation using maximum-likelihood method

To obtain the parameters of the BIN distribution using maximum-likelihood method, the likelihood function of Eq. (10) can be expressed as

$$L = \left(\frac{4}{\Gamma(a_1)\Gamma(a_2)}\left(\frac{a_1}{b_1}\right)^{a_1}\left(\frac{a_2}{b_2}\right)^{a_2}\right)^n \prod_{i=1}^{n}\left(t_{1i}^{-2a_1-1} t_{2i}^{-2a_2-1}\right)$$
$$\times \exp\left(-\sum_{i=1}^{n}\left(\frac{a_1}{b_1 t_{1i}^2}\right)\right)\exp\left(-\sum_{i=1}^{n}\left(\frac{a_2}{b_2 t_{2i}^2}\right)\right)$$
$$\times \prod_{i=1}^{n}\left\{1+\alpha\left(1-2\left(1-\gamma_1\left(a_1,\frac{a_1}{b_1 t_{1i}^2}\right)\right)\right)\left(1-2\left(1-\gamma_2\left(a_2,\frac{a_2}{b_2 t_{2i}^2}\right)\right)\right)\right\} \tag{11}$$

The log-likelihood function corresponding to Eq. (11) can be presented as

$$\ell = n\log\left(\frac{4}{\Gamma(a_1)\Gamma(a_2)}\left(\frac{a_1}{b_1}\right)^{a_1}\left(\frac{a_2}{b_2}\right)^{a_2}\right) - (2a_1+1)\sum_{i=1}^{n}(t_{1i}) - (2a_2+1)\sum_{i=1}^{n}(t_{2i}) \tag{12}$$
$$-\frac{a_1}{b_1}\sum_{i=1}^{n}\left(\frac{1}{t_{1i}^2}\right) - \frac{a_2}{b_2}\sum_{i=1}^{n}\left(\frac{1}{t_{2i}^2}\right) + \sum_{i=1}^{n}\log$$
$$\left(1+\alpha\left(1-2\left(1-\gamma_1\left(a_1,\frac{a_1}{b_1 t_{1i}^2}\right)\right)\right)\left(1-2\left(1-\gamma_2\left(a_2,\frac{a_2}{b_2 t_{2i}^2}\right)\right)\right)\right)$$

Now, we can derive the parameters of bivariate-inverted Nakagami distribution by differentiating Eq. (12) partially with respect to parameters $a_1, a_2, b_1, b_2,$ and α obtained as

$$\frac{\partial \ell}{\partial a_1} = -n\psi(a_1) + n\left(1+\log\left(\frac{a_1}{b_1}\right)\right) - 2\sum_{i=1}^{n}\log(t_{1i}) - \frac{1}{b_1}\sum_{i=1}^{n}\left(\frac{1}{t_{1i}^2}\right) \tag{13}$$
$$+\sum_{i=1}^{n}\left(\frac{\frac{\partial}{\partial a_1}(1+\alpha M_1 M_2)}{1+\alpha M_1 M_2}\right)$$

$$\frac{\partial \ell}{\partial a_2} = -n\psi(a_2) + n\left(1+\log\left(\frac{a_2}{b_2}\right)\right) - 2\sum_{i=1}^{n}\log(t_{2i}) - \frac{1}{b_2}\sum_{i=1}^{n}\left(\frac{1}{t_{2i}^2}\right) \tag{14}$$
$$+\sum_{i=1}^{n}\left(\frac{\frac{\partial}{\partial a_2}(1+\alpha M_1 M_2)}{1+\alpha M_1 M_2}\right)$$

$$\frac{\partial \ell}{\partial b_1} = -\frac{n a_1}{b_1} + \frac{a_1}{b_1^2}\sum_{i=1}^{n}\left(\frac{1}{t_{1i}^2}\right) + \sum_{i=1}^{n}\left(\frac{\frac{\partial}{\partial b_1}(1+\alpha M_1 M_2)}{1+\alpha M_1 M_2}\right) \tag{15}$$

$$\frac{\partial \ell}{\partial b_2} = -\frac{na_2}{b_2} + \frac{a_2}{b_2{}^2}\sum_{i=1}^{n}\left(\frac{1}{t_{2i}{}^2}\right) + \sum_{i=1}^{n}\left(\frac{\frac{\partial}{\partial b_2}(1+\alpha M_1 M_2)}{1+\alpha M_1 M_2}\right) \tag{16}$$

$$\frac{\partial \ell}{\partial \alpha} = -\sum_{i=1}^{n}\left(\frac{1}{1+\alpha M_1 M_2}\right)\frac{\partial}{\partial \alpha}(1+\alpha M_1 M_2) \tag{17}$$

Simplifying Eqs. (13)–(17) and then equating to zero will yield the estimates of the parameters of the bivariate-inverted Nakagami distribution.

4.2 Estimation using inference functions for margins

Estimation using inference function for margins can be obtained by considering marginal density functions of the bivariate-inverted Nakagami distribution. The marginal density functions of the bivariate Maxwell distribution can be derived as:

4.2.1 Marginal density function of T_1

The marginal density function of T_1 can be defined as

$$f_1(t_1) = \int_{\infty}^{\infty} f(t_1, t_2)dt_2 \tag{18}$$

where $f(t_1, t_2)$ is defined in Eq. (10). Substituting Eq. (10) into Eq. (18) gives

$$f_1(t_1) = \frac{4}{\Gamma(a_1)\Gamma(a_2)}\left(\frac{a_1}{b_1}\right)^{a_1}\left(\frac{a_2}{b_2}\right)^{a_2}\int_0^{\infty} t_1{}^{-2a_1-1}t_2{}^{-2a_2-1}\exp\left(-\frac{a_1}{b_1 t_1{}^2}\right)\exp\left(-\frac{a_2}{b_2 t_2{}^2}\right)$$
$$\left\{1+\alpha\left(1-2\left(1-\gamma_1\left(a_1,\frac{a_1}{b_1 t_1{}^2}\right)\right)\right)\left(1-2\left(1-\gamma_2\left(a_2,\frac{a_2}{b_2 t_2{}^2}\right)\right)\right)\right\}dt_2 \tag{19}$$

Let

$$A = 1-\gamma_2\left(a_2,\frac{a_2}{b_2 t_2{}^2}\right), \quad \Rightarrow \quad dt_2 = \frac{\Gamma(a_2)b_2{}^{a_2}t_2{}^{2a_2+1}}{2a_2{}^{a_2}e^{-\frac{a_2}{b_2 t_2{}^2}}}dA \tag{20}$$

Inserting Eq. (20) into Eq. (19) becomes

$$f_1(t_1) = \frac{2}{\Gamma(a_1)}\left(\frac{a_1}{b_1}\right)^{a_1} t_1{}^{-2a_1-1}\exp\left(-\frac{a_1}{b_1 t_1{}^2}\right)\int_0^1\left\{1+\alpha\left(1-2\left(1-\gamma_1\left(a_1,\frac{a_1}{b_1 t_1{}^2}\right)\right)\right)(1-2A)\right\}dA$$

$$= f(t_1;a_1,b_1) + \alpha\left(1-2\left(1-\gamma_1\left(a_1,\frac{a_1}{b_1 t_1{}^2}\right)\right)\right)f(t_1;a_1,b_1)\int_0^1\{(1-2A)\}dA$$

$$= f(t_1;a_1,b_1) + \alpha\left(1-2\left(1-\gamma_1\left(a_1,\frac{a_1}{b_1 t_1{}^2}\right)\right)\right)f(t_1;a_1,b_1)\{0\}$$

$$f_1(t_1) = \frac{2}{\Gamma(a_1)}\left(\frac{a_1}{b_1}\right)^{a_1} t_1{}^{-2a_1-1}\exp\left(-\frac{a_1}{b_1 t_1{}^2}\right) \tag{21}$$

which is the marginal density function of T_1. Hence, the marginal density function of T_2 can be presented as

$$f_2(t_2) = \frac{2}{\Gamma(a_2)}\left(\frac{a_2}{b_2}\right)^{a_2} t_2^{-2a_2-1} \exp\left(-\frac{a_2}{b_2 t_2^2}\right) \quad (22)$$

The parameter estimations of the marginal densities of T_1 and T_2 using inference function for margins can be obtained from Eqs. (21) and (22), and then the log-likelihood function of these equations can be presented as

$$\ell_{T_{1i}} = n\log(2) - n\log(\Gamma(a_1)) + na_1\log(a_1) - na_1\log(b_1) - (2a_1+1)\sum_{i=1}^{n}\log(t_{1i}) - \frac{a_1}{b_1}\sum_{i=1}^{n}\left(\frac{1}{t_{1i}^2}\right) \quad (23)$$

and

$$\ell_{T_{2i}} = n\log(2) - n\log(\Gamma(a_2)) + na_2\log(a_2) - na_2\log(b_2) - (2a_2+1)\sum_{i=1}^{n}\log(t_{2i}) - \frac{a_2}{b_2}\sum_{i=1}^{n}\left(\frac{1}{t_{2i}^2}\right) \quad (24)$$

Maximizing Eqs. (23) and (24) over parameters a_j and b_j for $j = 1, 2$, and then equating to zero we can have.

$$\frac{\partial \ell_{T_{ji}}}{\partial a_j} = -n\psi(a_j) + n(1 + \log(a_j)) - n\log(b_j) - 2\sum_{i=1}^{n}(t_{ji}) - \frac{1}{b_j}\sum_{i=1}^{n}\left(\frac{1}{t_{ji}^2}\right) = 0 \quad (25)$$

Eq. (25) is nonlinear and it cannot be derived numerically, the statistical software such as R, MATLAB, and so on could be employed effectively in estimating the parameters of the marginal density functions of of T_1 and T_2 Furthermore, the parameter α in Eq. (10) can be obtained using the following ways:

$$\ell_\alpha = \sum_{i=1}^{n}\log\left(1 + \alpha\left(1 - 2\left(1 - \gamma_1\left(a_1, \frac{a_1}{b_1 t_{1i}^2}\right)\right)\right)\left(1 - 2\left(1 - \gamma_2\left(a_2, \frac{a_2}{b_2 t_{2i}^2}\right)\right)\right)\right) \quad (26)$$

The partial derivative with respect to parameter α in Eq. (26), and equating zero it becomes

$$\frac{\partial \ell_\alpha}{\partial \alpha} = \sum_{i=1}^{n}\left(\frac{\left(1 - 2\left(1 - \gamma_1\left(a_1, \frac{a_1}{b_1 t_{1i}^2}\right)\right)\right)\left(1 - 2\left(1 - \gamma_2\left(a_2, \frac{a_2}{b_2 t_{2i}^2}\right)\right)\right)}{1 + \alpha\left(1 - 2\left(1 - \gamma_1\left(a_1, \frac{a_1}{b_1 t_{1i}^2}\right)\right)\right)\left(1 - 2\left(1 - \gamma_2\left(a_2, \frac{a_2}{b_2 t_{2i}^2}\right)\right)\right)}\right) = 0 \quad (27)$$

equating (27), and then simplifying for α gives the estimate of the parameter of the bivariate-inverted Nakagami distribution using the maximum-likelihood estimation.

5. Application to real-life datasets

The effectiveness of the new bivariate-inverted Nakagami distribution based on the two datasets is evaluated through an application to real-world datasets. The first dataset for kidney infection diseases is given in [48] and involves 30 observations. The second dataset for the UEFA Champions' League group stage for the seasons 2004–2005 and 2005–2006 is presented [49].

Table 1 presents the mean estimate (estimate), standard error (std. error), T-square (t) and probability (p) values for the kidney infection diseases and the group stage of the UEFA Champion's League.

The estimates, standard error, t, and p values for the MLE and IFM approaches are shown in **Table 1**. Based on standard error values, the IFM estimates of the parameters are generally superior to the corresponding MLE estimates. With the exception of the parameters b_2 and α from the MLE and IFM, respectively, both estimation techniques are statistically significant. This determines whether the FGM copula is appropriate for the first dataset.

Table 2 shows that the findings of the IFM are superior to those of the MLE, and the corresponding p values for each parameter are statistically significant as well. This demonstrates that the FGM copula is appropriate for the second datasets.

The copula goodness of fit measures for kidney disease infection and the group stage of the UEFA Champions League are presented in **Tables 3** and **4**. The copula goodness and fit measures of the IMF and MLE technique of estimations are measured by the log-likelihood (log-lik), Akaike Information Criterion (AIC), Bayesian Information Criterion (BIC), and Hannan-Quinn Information Criterion (HQIC). The best technique should be defined as having a maximum log-lik value and a minimum AIC, BIC, and HQIC value.

Method	Parameter	Estimate	Std. error	t	p
	a_1	0.2041	0.0278	7.4060	0.0000
	b_1	1.3083	0.5684	2.3020	0.0214
MLE	a_2	0.2654	0.0499	5.3130	0.0000
	b_2	0.2234	0.1581	1.4130	0.1576
	α	10.6062	4.6728	2.2700	0.0232
	a_1	0.2251	0.0449	5.0090	0.0000
	b_1	0.0122	0.0050	2.4250	0.0153
IFM	a_2	0.2916	0.0593	4.9150	0.0000
	b_2	0.0042	0.0015	2.8300	0.0047
	α	1.0586	0.6184	1.7120	0.0869

Table 1.
Goodness-of-fit measures for the kidney infection diseases.

Method	Parameter	Estimate	Std. error	t	p
	a_1	0.2506	0.0323	7.7470	0.0000
	b_1	0.0084	0.0021	3.9820	0.0000
MLE	a_2	0.3387	0.0601	5.6370	0.0000
	b_2	0.0217	0.0092	2.3630	0.0181
	α	3.4233	1.1019	3.1070	0.0018
	a_1	0.3085	0.0568	5.4350	0.0000
	b_1	0.0079	0.0023	3.4660	0.0005
IFM	a_2	0.3202	0.0591	5.4180	0.0000
	b_2	0.0161	0.0048	3.3350	0.0009
	α	2.3231	0.6054	3.8370	0.0001

Table 2.
Goodness of fit measures for the group stage of the UEFA Champion's league.

Copula	Log-lik	AIC	BIC	HQIC
IFM	−348.2453	706.4906	713.4966	708.7319
MLE	−350.8327	711.6654	718.6714	713.9067

Table 3.
Copula goodness-of-fit measures results for the first dataset.

Copula	Log-lik	AIC	BIC	HQIC
IFM	−371.2927	752.5854	760.6400	755.4250
MLE	−371.4730	752.9460	761.0006	755.7856

Table 4.
Copula goodness-of-fit measures results for the second dataset.

Table 3 shows that the IFM provides a minimal value for the AIC, BIC, and HQIC and a maximum value for log-lik. This proved that the IMF's approach to finding the bivariate-inverted Nakagami distribution's parameters was superior.

Table 4 shows the findings of the IFM and MLE, and it is evident from this table that the estimation using the IFM gave the best results, having a higher log-lik value and with the smallest AIC, BIC, and HQIC values. **Table 4** shows the findings of the IFM and MLE, and it is evident from this table that the estimation using the IFM gave the best results, having a higher log-lik value and with the smallest AIC, BIC, and HQIC values.

6. Conclusion

This chapter introduces a brand-new bivariate inverted Nakagami distribution, along with its characteristics and practical applications. The new bivariate distribution's cdf, pdf, and marginal density functions are specified. The model parametrs were estimated using a variety of estimation techniques. To demonstrate the effectiveness of the novel distribution, two datasets are taken into account. The findings indicate that the IFM produced the most accurate method for estimating the bivariate-inverted Nakagami distribution's parameters.

Acknowledgements

This research was funded by the National Collaborative Research Fund (015MC0-032). The Universiti Teknologi PETRONAS is acknowledged by authors for providing facilities to this project.

Conflict of interest

The authors claim to have no conflicts of interest.

Declarations

We certify that all authors have reviewed and approved this chapter.

Author details

Aliyu Ismail Ishaq[1*], Abubakar Usman[1], Ahmad Abubakar Suleiman[2,3], Mahmod Othman[2], Hanita Daud[2], Rajalingam Sokkalingam[2], Uthumporn Panitanarak[4] and Muhammad Azrin Ahmad[5]

1 Department of Statistics, Ahmadu Bello University, Zaria, Nigeria

2 Fundamental and Applied Sciences Department, Universiti Teknologi Petronas Seri Iskandar, Malaysia

3 Department of Statistics, Aliko Dangote University of Science and Technology, Wudil, Nigeria

4 Department of Public Health, Mahidol University, Thailand

5 Centre for Mathematical Sciences, Universiti Malaysia, Pahang, Malaysia

*Address all correspondence to: binishaq05@gmail.com

References

[1] Nakagami M. The m-distribution—A general formula of intensity distribution of rapid fading. In: Statistical Methods in Radio Wave Propagation. In: Proceedings of a Symposium Held at the University of California, Los Angeles, June 18–20, 1958. Elsevier; 1960. pp. 3-36. DOI: 10.1016/B978-0-08-009306-2.50005-4

[2] Cui W et al. Automatic segmentation of ultrasound images using SegNet and local Nakagami distribution fitting model. Biomedical Signal Processing and Control. 2023;**81**:104431. DOI: 10.1016/j.bspc.2022.104431

[3] Liu Z, Du Y, Meng X, Li C, Zhou L. Temperature monitoring during microwave hyperthermia based on BP-Nakagami distribution. Journal of Ultrasound in Medicine. 2023. DOI: 10.1002/jum.16213

[4] Rathnam MJ, Christ J. A novel method for cataract detection and segmentation using Nakagami distribution. Journal of Medical Imaging and Health Informatics. 2022; **12**(1):45-51

[5] Pajala E, Isotalo T, Lakhzouri A, Lohan ES, Renfors M. An improved simulation model for Nakagami-m fading channels for satellite positioning applications. In: 3rd Workshop on Position, Navigation and Communication. Hannover, Germany: Academia; 2006. pp. 81-89

[6] Reig J. Multivariate Nakagami-m distribution with constant correlation model. AEU-International Journal of Electronics and Communications. 2009; **63**(1):46-51

[7] Simon MK, Alouini M-S. A unified performance analysis of digital communication with dual selective combining diversity over correlated Rayleigh and Nakagami-m fading channels. IEEE Transactions on Communications. 1999;**47**(1):33-43

[8] Zhang QT. Maximal-ratio combining over Nakagami fading channels with an arbitrary branch covariance matrix. IEEE Transactions on Vehicular Technology. 1999;**48**(4): 1141-1150

[9] Beaulieu NC, Cheng C. Efficient Nakagami-m fading channel simulation. IEEE Transactions on Vehicular Technology. 2005;**54**(2):413-424

[10] Alavi O, Mohammadi K, Mostafaeipour A. Evaluating the suitability of wind speed probability distribution models: A case of study of east and southeast parts of Iran. Energy Conversion and Management. 2016;**119**: 101-108. DOI: 10.1016/j.enconman.2016.04.039

[11] Datta P, Gupta A, Agrawal R. Statistical modeling of B-mode clinical kidney images. In: 2014 International Conference on Medical Imaging, M-Health and Emerging Communication Systems (MedCom). Noida, India: IEEE; 2014. pp. 222-229. DOI: 10.1109/MedCom.2014.7006008

[12] Zhou Z, Wu S, Wang C-Y, Ma H-Y, Lin C-C, Tsui P-H. Monitoring radiofrequency ablation using real-time ultrasound Nakagami imaging combined with frequency and temporal compounding techniques. PLoS One. 2015;**10**(2):e0118030

[13] Sarkar S, Goel NK, Mathur B. Adequacy of Nakagami-m distribution function to derive GIUH. Journal of Hydrologic Engineering. 2009;**14**(10): 1070-1079

[14] Sarkar S, Goel NK, Mathur B. Performance investigation of Nakagami-m distribution to derive flood hydrograph by genetic algorithm optimization approach. Journal of Hydrologic Engineering. 2010;**15**(8):658-666

[15] Rai R, Sarkar S, Upadhyay A, Singh V. Efficacy of Nakagami-m distribution function for deriving unit hydrograph. Water Resources Management. 2010;**24**:563-575

[16] Shankar PM et al. Classification of ultrasonic B-mode images of breast masses using Nakagami distribution. IEEE Transactions on Ultrasonics, Ferroelectrics, and Frequency Control. 2001;**48**(2):569-580

[17] Tsui P-H, Huang C-C, Wang S-H. Use of Nakagami distribution and logarithmic compression in ultrasonic tissue characterization. Journal of Medical and Biological Engineering. 2006;**26**(2):69-73

[18] Nakahara H, Carcolé E. Maximum-likelihood method for estimating coda Q and the Nakagami-m parameter. Bulletin of the Seismological Society of America. 2010;**100**(6):3174-3182

[19] Ahmad K, Ahmad S, Ahmed A. Classical and Bayesian approach in estimation of scale parameter of Nakagami distribution. Journal of Probability and Statistics. 2016;**2016**: 7581918. DOI: 10.1155/2016/7581918

[20] Zhang S et al. Feasibility of using Nakagami distribution in evaluating the formation of ultrasound-induced thermal lesions. The Journal of the Acoustical Society of America. 2012; **131**(6):4836-4844. DOI: 10.1121/ 1.4711005

[21] Schwartz J, Godwin RT, Giles DE. Improved maximum-likelihood estimation of the shape parameter in the Nakagami distribution. Journal of Statistical Computation and Simulation. 2013;**83**(3):434-445. DOI: 10.1080/ 00949655.2011.615316

[22] Huang L-F. The Nakagami and its related distributions. WSEAS Transactions on Mathematics. 2016; **15**(44):477-485

[23] Zaka A, Akhter AS. Bayesian approach in estimation of scale parameter of Nakagami distribution. Pakistan Journal of Statistics and Operation Research. 2014;**10**(2):217-228

[24] Artyushenko VM, Volovach VI. Nakagami distribution parameters comparatively estimated by the moment and maximum likelihood methods. Optoelectronics, Instrumentation and Data Processing. 2019;**55**(3): 237-242. DOI: 10.3103/S87566990190 3004X

[25] Kumar K, Garg R, Krishna H. Nakagami distribution as a reliability model under progressive censoring. International Journal of System Assurance Engineering and Management. 2017;**8**(1):109-122. DOI: 10.1007/s13198-016-0494-3

[26] Ahad SM, Ahmad SP. Characterization and estimation of the length biased Nakagami distribution. Pakistan Journal of Statistics and Operation Research. 2018;**14**(3):697-715

[27] Abdullahi I, Phaphan W. A generalization of length-biased Nakagami distribution. International Journal of Mathematics andComputer Science. 2022;**17**:21-31

[28] Louzada F, Ramos PL, Nascimento D. The inverse Nakagami-m distribution: A novel approach in

reliability. IEEE Transactions on Reliability. 2018;**67**(3):1030-1042

[29] Morgenstern D. Einfache beispiele zweidimensionaler verteilungen. Mitteilingsblatt fur Mathematische Statistik. 1956;**8**:234-235

[30] Gumbel EJ. Bivariate logistic distributions. Journal of the American Statistical Association. 1961;**56**(294): 335-349

[31] Farlie DJ. The performance of some correlation coefficients for a general bivariate distribution. Biometrika. 1960; **47**(3/4):307-323

[32] Johnson NL, Kott S. On some generalized farlie-gumbel-morgenstern distributions. Communications in Statistics-Theory and Methods. 1975; **4**(5):415-427

[33] Marshall AW, Olkin I. A multivariate exponential distribution. Journal of the American Statistical Association. 1967; **62**(317):30-44

[34] Mohammed BI, Hossain MM, Aldallal RA, Mohamed MS. Bivariate Kumaraswamy distribution based on conditional hazard functions: Properties and application. Mathematical Problems in Engineering. 2022;**2022**:2609042. DOI: 10.1155/2022/2609042

[35] Mohammed B, Makumi N, Aldallal R, Dyhoum TE, Aljohani HM. A new model of discrete-continuous bivariate distribution with applications to medical data. Computational and Mathematical Methods in Medicine. 2022;**2022**

[36] Alotaibi R, Nassar M, Ghosh I, Rezk H, Elshahhat A. Inferences of a mixture bivariate alpha power exponential model with engineering application. Axioms. 2022;**11**(9):459 Available from: https://www.mdpi.com/2075-1680/11/9/459

[37] Ali MM, Mikhail N, Haq MS. A class of bivariate distributions including the bivariate logistic. Journal of Multivariate Analysis. 1978;**8**(3):405-412

[38] Reig J, Rubio Arjona L, Cardona Marcet N. Bivariate Nakagami-m distribution with arbitrary fading parameters. Electronics Letters. 2002; **38**(25):1715-1717

[39] Souza RAAD, Yacoub MD. Bivariate Nakagami-m distribution with arbitrary correlation and fading parameters. IEEE Transactions on Wireless Communications. 2008;**7**(12):5227-5232. DOI: 10.1109/T-WC.2008.071152

[40] Souza RAAD, Yacoub MD. On the multivariate Nakagami-m distribution with arbitrary correlation and fading parameters. In: 2007 SBMO/IEEE MTT-S International Microwave and Optoelectronics Conference, Salvador, Brazil: IEEE; Oct.-1 Nov. 2007. pp. 812-816. DOI: 10.1109/ IMOC.2007.4404382

[41] Yang J, Qi Y, Wang R. A class of multivariate copulas with bivariate Fréchet marginal copulas. Insurance: Mathematics and Economics. 2009; **45**(1):139-147

[42] Myrhaug D, Leira B. A bivariate Fréchet distribution and its application to the statistics of two successive surf parameters. Proceedings of the Institution of Mechanical Engineers, Part M: Journal of Engineering for the Maritime Environment. 2011;**225**(1): 67-74

[43] Zheng Y, Yang J, Huang JZ. Approximation of bivariate copulas by patched bivariate Fréchet copulas.

Insurance: Mathematics and Economics. 2011;**48**(2):246-256

[44] Nelsen RB. An Introduction to Copulas. New York. MR2197664: Springer; 2006

[45] Sklar A. Random variables, joint distribution functions, and copulas. Kybernetika. 1973;**9**(6):449-460

[46] Gumbel EJ. Bivariate exponential distributions. Journal of the American Statistical Association. 1960;**55**(292): 698-707

[47] Almetwally EM, Muhammed HZ, El-Sherpieny E-SA. Bivariate Weibull distribution: Properties and different methods of estimation. Annals of Data Science. 2020;7:163-193

[48] Kumar P. Probability distributions and estimation of Ali-Mikhail-Haq copula. Applied Mathematical Sciences. 2010;**4**(14):657-666

[49] Almetwally EM, Muhammed HZ. On a bivariate Fréchet distribution. Journal of Statistics Applications & Probability. 2020;**9**(1):1-21

[50] Plackett RL. A class of bivariate distributions. Journal of the American Statistical Association. 1965;**60**(310): 516-522. DOI: 10.1080/01621459. 1965.10480807

Chapter 6

Perspective Chapter: Science and Technology Libraries in the Age of Open Science – Scenarios for the New Protagonism of Scientific and Technological Information

Lillian Alvares and Kira Tarapanoff

Abstract

Science and Technology Libraries (STL) have always functioned as supporters of scientific and technological activities. In the Digital Era, this role was considered superfluous due to the facilities offered by information and communication technologies. In this work, we consider that the recent movements of Open Science and open access to scientific publications enable libraries to become again important protagonists in the scientific community. In this context, our objective is to analyze the relationship between the library and the Open Science proposal considering the complex elements that constitute the value chain of scientific and technological research. Aspects of analysis include collaboration, multilevel governance, co-production, and co-creation, with the pragmatic envelopment of information. The chosen method is the philosophical current of critical realism presenting a conceptual framework that relates STL, Open Science, organizational collaboration, multilevel governance, and current scientific information.

Keywords: science and technology libraries (STL), Open Science, collaboration, multilevel governance, current research information system (CRIS)

1. Introduction

Science and Technology Libraries (STL) are defined as libraries that support scientific and technological research wherever it takes place. Nevertheless, the advancement of information and communication technologies has somewhat diminished the central role they once held in the research chain. In the last 20 years, however, the Open Science movement has emerged and, in particular, the segments of open access to scientific publications and open research data have become significant contributions that libraries can offer to the scientific community. In this context, the objective of this chapter is to bring the relationship between the library and Open

Science and to add to the implex elements that support the perennial protagonism of the library in the value chain of scientific and technological research, such as collaboration, multilevel governance, co-production and co-creation, and the pragmatic wrapping of current information. The methodology adopted is founded in the philosophical current of critical realism, applied, qualitative, exploratory, longitudinal, and bibliographic, opting as methodological strategy the narrative literature review, followed by qualitative and inductive content analysis. The result presented is a conceptual framework that relates STL, Open Science, organizational collaboration, multilevel governance, and current scientific information. In conclusion, the chapter shows that the discussion about the role of libraries in science and technology does not end with the philosophical issues of Open Science and the competencies of information science and computer science. It moves on to political and organizational issues since Open Science involves an intense process of governance and collaboration.

Wilson [1] defines STL with words that transcend the physical infrastructure available, *"as the source of ideas and ideals and as the stimulator of scholarly interests and attitudes"* (p. 144). They were the center of scientific and technological research, represented by imposing buildings in universities and research centers when their definition was *"a place where the scholar, beginning with a thesis or with a question, can pursue it wherever it leads … allows the researcher to follow – with efficiency – an idea that spontaneously arises, it also permits accidental discovery"* (Fabian [2] apud [3]).

However, the evolution of information and communication technologies removed the leading role of research from libraries and gave researchers the autonomy to go directly to the digital space of scientific and technological resources and information. For a while, the main source of access and dissemination of advances in science and technology was not at the forefront of efforts to advance knowledge.

In the last 20 years came in force the Open Science movement, an initiative that has been mobilizing the science and technology community. This movement is strengthening the ethos of science described by Robert K. Merton, since for this sociologist; the search for the common good must be the ontology that governs the behavior of the researcher. In the following decade, two other manifestations illuminated what would become Open Science.

Article 27 of the Declaration of Human Rights, which guarantees the right of all man to participate in scientific progress and its benefits, and the publication of the book Selected Poems, by Mark Van Doren, winner of the Pulitzer Prize for Poetry, who defends everyone's right to the free use of knowledge.

In the context of the role of STL and the flourishing of Open Science, this research aims to investigate the relationship between the library and Open Science and to add to the implicit elements that support the role of the library in the value chain of scientific and technological research. Specifically, it intends to present the simplified scenario of Open Science (contemporary meanings based on history, schools of thought, and the positioning of international organizations). In addition, defining aspects of Open Science in the research library as its structuring elements include collaboration and multilevel governance, the paradigm of co-production, and co-creation.

Regarding the methodology, the research is grounded in Roy Bhaskar's (1944–2014) philosophical approach of critical realism, which considers social life to be an open system constructed by several dimensions, each with its distinctive structures [4]. This research can be considered applied science character from the point of view of its nature because its goal is to generate new knowledge for practical application. The approach to the problem is qualitative. From the point of view of the objectives, it is a secondary exploratory and propositional research. From the time perspective, it is longitudinal and from the point of view of technical procedures, it is bibliographical research. The methodological strategy is a narrative literature review, based on the consensus and criticism of the available scientific production. The selected studies are, essentially, fundamental epistemological works in each of the thematic cores of the research (Open Science, science, technology libraries, organizational collaboration, multilevel governance, and current scientific information systems). After the review, the next procedure was inductive content analysis. It consisted of an in-depth analysis of selected works based on the methodology proposed by Elo and Kyngäs [5], recommended when there are no previous studies dealing with the phenomenon or when knowledge is fragmented.

The expected result is a conceptual framework that relates STL, Open Science, organizational collaboration, multilevel governance, and current scientific information, to understand the structuring condition of the elements mentioned as the set of intra and inter-organizational, interconnected, and interdependent activities, which add value to scientific and technological research.

2. Open Science

2.1 Contemporary meanings historically grounded

The term Open Science, with its current meaning, originates in the 1985 paper Open science and closed science: tradeoffs in a democracy, written by Daryl E. Chubin, in which the author enters into Robert Merton's 1942 classic on the imperatives for scientific practice—communism, universalism, communication, disinterestedness, and organized skepticism—presenting the idiosyncrasies related to the social context to enable the ethos of Open Science.

Thirty years later, Vercellone et al. [6] reinforced the spirit of the common good that innervates the open nature of information and communication technologies in meeting the Mertonian culture and ethos of Open Science, with the following words: "*the new generation brought up on widespread knowledge takes up and reformulates the four fundamental Mertonian principles of universalism, communism, disinterestedness, and organized skepticism on their own, integrating them into a new value system ...* " (p. 58).

In that paper, Chubin [7] also defines closed science as "*research which, in its production, communication, or utilization, is inaccessible to potential consumers. The grounds for such closure are always political*" (p. 80). And continues: "*the contrasts between Open Science and closed science help to clarify, much like the norms, how constituent communities of science adapt their behavior to ideology on the one hand and local organizational constraints on the other .... It is a matter for political debate, not scientific judgment alone*" (p. 81).

Collaboration and participation, on the other hand, are an integral part of the Open Science ecosystem. Its actors collectively strive in the pursuit and evolution of the set of governance norms for the production and dissemination of reliable knowledge. According to ref. [8], the predicate open makes precise and robust the meaning of promoting a common good, in which participation is welcome and interoperability is maximized: "*knowledge is open if anyone is free to access, use, modify, and share it — subject, at most, to measures that preserve provenance and openness*" (n.d.).

Reference [9] includes noting that the results of Open Science are subject to open copyright, that on a legal level allows the use, modification, and transfer of knowledge, in this case scientifically proven knowledge[1]. Burgelman et al. [10] claim that *"it is therefore very likely that in the long term, the adjective open should not be necessary as science will be open by default"* (p. 1). Even if governments claim so, Open Science and open access will become the norm in academic research [11]. Meanwhile, according to ref. [12], the openness of science oscillates toward a workable combination of availability, legality, and costs.

Open Science revolves around the fundamental principles of utilizing, reusing, and sharing knowledge, all facilitated by open digital technologies that guarantee the realization of this ideal. From an economic point of view and by a functionalist explanation, Open Science can be justified by full access to data and information throughout the research cycle, reduced duplication of research effort, rapid validation of findings, expanded knowledge, and cooperation between research initiatives. It certainly is that, but it goes further. By considering Open Science a common good, the movement makes scientific research accessible at all levels of society, facilitating enjoyment, encouraging engagement, and accelerating new practices and achievements.

Roughly speaking, Open Science drives and is related to open access publication, open data, open educational resources, open engagement of social actors (citizen science), open evaluation (or open peer review), open hardware, open innovation, Open Science infrastructures, open source software, openness to all scholarly knowledge and inquiry, openness to diversity knowledge, and openness to indigenous systems. Initiatives that can be considered the turning point for new science, supported by technologies that multiply the impact of the results in the very ecosystem of Open Science and its transformative potential for reducing social inequalities and accelerating the progress of humanity (**Figure 1**).

From an information science perspective, open access to scientific publications and open research data are the main movements of Open Science. The first, established in

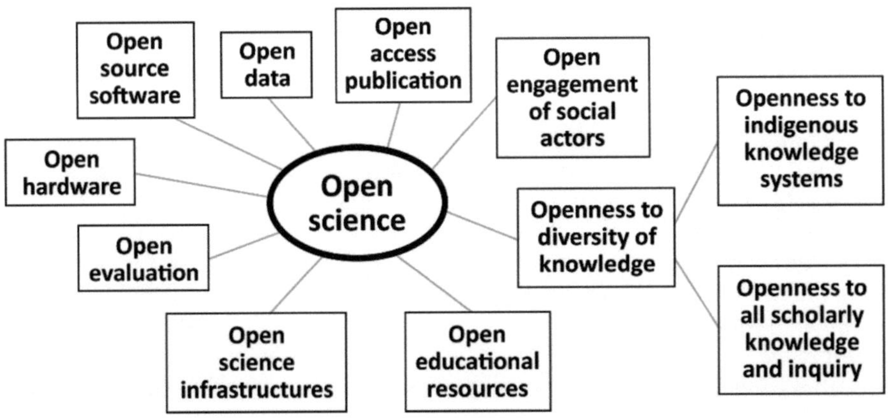

Figure 1.
Open Science diagram from UNESCO (source: Persic [13]).

[1] The European Commission's report Consultation on 'Science 2.0': Science in Transition carried out a public consultation between July and September 2014 and defined the term Open Science as preferred among six options (among them Science 2.0 and e-Science) (European Commission, 2014).

2002 within the scope of The Budapest Open Access Initiative (BOAI), formalized the Open Access Movement[2] and coined the term open access.

The declaration made the model of "*eliminating barriers that prevent the legitimate use of scientific literature for academic purposes [...] available online, without economic barriers and most reuse permission barriers*[3]"([14], p. 63)[4]. The meaning and technologies associated with the open access of scientific publications led not only to open access but also to free dissemination, based on licenses such as the Creative Commons in 2002. According to ref. [15], "*Open Access to research results is an essential part of Open Science, which aims to make science more reliable, efficient and responsive*" (p. 15). About Open data, Section 2.2 will deal with this topic in depth.

And the second, open research data, is one of the most important elements for the success of Open Science [10]. With them it is possible to reuse and enrich the dataset, the result of which is to shorten the time needed for research. It is worth considering that through open research data, it is possible to increase the critical evaluation of research, detecting inaccuracies and inaccuracies and enabling more accurate replicability tests.

It is important to note that Open Science is also concerned with crediting the ownership of research efforts to the respective authors, and this extends to research data as well, giving visibility to those responsible for collecting or generating the data, increasing their citations, and therefore increasing their research impact index.

As more researchers have adopted Open Science practices, allowing wider sharing of research results, research data has gradually become the focus of attention in STL. According to Federer et al. [16], many STL offer research data management (RDM), with an emphasis on data management planning.

However, it should not be lost sight that Open Science is related to the entire research process (collecting, analyzing, publishing, reanalyzing, criticizing, and reusing), and "*represents a new approach to the scientific process based on cooperative work and new ways of diffusing knowledge by using digital technologies and new collaborative tools*" ([15], p. 33). Indeed, the OECD Blue Sky III Forum, Smith et al. [17] defined Open Science as openness to the entire research cycle referring to *ongoing changes in the way research is conducted, with a move toward increased transparency, collaboration, communication, and participation*" (p. 1) (**Figure 2**).

To facilitate the understanding and quick visualization of all the segments reached by Open Science, Pontika et al. [18] made available the Open Science Taxonomy, a collaboratively developed mind map that establishes the hyponymic relationship of the area. The reason for creating the taxonomy was to represent the concepts of this special purpose vocabulary, in an organized way [19]. It provides standardization of the ways of expression, avoiding ambiguities and improving the quality of communication of terms and concepts of this specialized language. The main issues in creating the taxonomy were the joint decisions about the number of representative hierarchies to be chosen, the terminology that should be adopted, and the relationship between the terms. With this orientation, nine terms were selected to represent the first hierarchical level, with the ramifications shown in **Figure 3**.

---

[2] The Open Access milestone is in the declarations of Budapest (February 2002), Bethesda (June 2003), and Berlin (October 2003), also known as BBB.

[3] Eliminación de las barreras que impidan el uso legítimo de la literatura científica con fines académicos ... disponible online, sin barreras económicas y sin la mayoría de las barreras de los permisos de reutilización.

[4] Here in the 2015 edition in Spanish, p. 63–64.

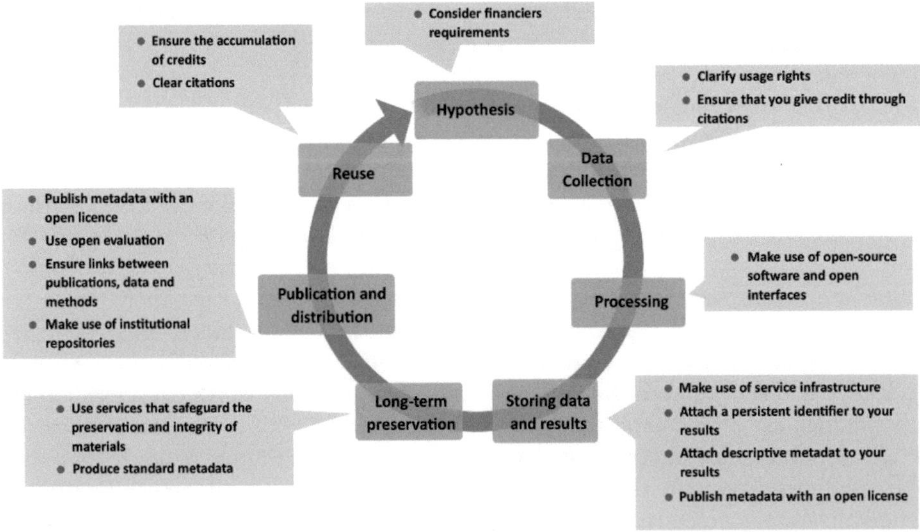

Figure 2.
Different stages of research and the corresponding opening processes (Source: Finnish Open Science and Research Initiative [12]).

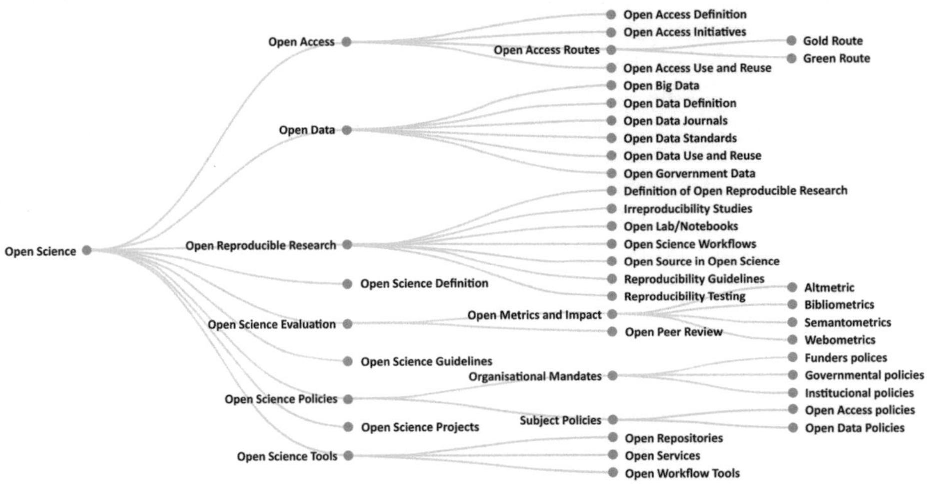

Figure 3.
Open Science taxonomy (source: Knoth and Pontika [20]).

2.2 Open Science schools of thought

With the diversity of approaches to Open Science, sociologists [21] structured and synthesized five schools of thought, which describe the different interpretations of the term and its principles.

2.2.1 Infrastructure school

Centered on the creation of available platforms, tools, and services, promoting the infrastructure for the development of research. Approaches Open Science as a technological challenge, represented in two trends. The first is distributed computing, which allows high-volume data processing in large projects, from computer networks and technologies around the world [22]. The second is social networking based on the collaboration of scientists, with technological resources that facilitate the collaboration and definition of the social virtual research environment. Characterized by:

i. Sharing resources frequently used by researchers;

ii. Providing incentives for researchers to make their research objects available on the platform;

iii. Keeping the digital artifacts that make up the environment easily integrable and;

iv. Preparing the environment, not only, to be a place to store information and research resources, but also to be used for conducting research [23].

2.2.2 Measurement school

The authors, of this school, point out that measuring the scientific impact of scientific contribution is capital for the researcher's career and research. They argue that, in effect, Open Science revolves around the development of robust alternative methods of measuring science, not least because of criticism of the prevailing modus operandi. Contemporary technological conditions are sufficient to obtain a full multi-faceted and multidimensional assessment of the resulting impact, which is under the hypernym of altmetrics, a term coined by Priem et al. [24], which refers *"to study and use of scholarly impact measures based on activity in online tools and environments. (...) is in most cases a subset of both scientometrics and webometrics"* ([25], p. e48753).

2.2.3 Public school

This understanding is concerned with making science accessible to a wider public, considers two perspectives: in the accessibility of the research process and; in the understanding of the research results for the society and not only for the specialists of the ecosystem directly involved in the production and fruition. Access to the public is about promoting the inclusion of individuals external to the research process. Its objective is to expand the reach of research, the multiplication of the data examined, the collective intelligence of the volunteer workforce, confirming what is being called citizen science, initiatives that include the participation of amateurs, volunteers, and enthusiasts in large-scale scientific projects, usually limited to data collection [26]. Making research understandable to a wider public, including the fundamental concept that every researcher should make their research accessible to the public, also considers science popularization initiatives.

2.2.4 Democratic school

Here are concentrated initiatives of open access to scientific publications and open research data, which are concerned with access to knowledge, of the products of research, trying to solve the legal obstacles that hinder access to research publications and scientific data. It meets the democratic principle, advocated in the Declaration of

Human Rights, that knowledge should be available to all and that all have an equal right to access to knowledge, especially when it comes to publicly funded research. The Universal Declaration of Human Rights, adopted by the United Nations Assembly in 1948, states in Article 27 that "*1. Everyone has the right freely to participate in the cultural life of the community, to enjoy the arts and to share in scientific advancement and its benefits. 2. Everyone has the right to the protection of the moral and material interests resulting from any scientific, literary or artistic production of which he is the author*" [27].

2.2.5 Pragmatic school

In this school of thought, the core is collaboration and its intrinsic benefits in dealing with critical issues, such as the creation and dissemination of new knowledge from scientific research. This is, lato sensu, open innovation, which: "*adopts an integrative perspective when considering internal and external sources of knowledge, organizations must acquire knowledge from external actors to integrate with internally developed knowledge, which is in line with West and Bogers (2014)* [28] *who present the issue as summon to people within organizations to seek external knowledge and integrate it with internal knowledge to improve processes and products*" (p. 121).

In summary, Open Science brings opportunities to increase collaboration throughout the research process. As ref. (21) summarize by Tacke's [29] thinking on open innovation, Open Science transfers "*from the outside-in (including external knowledge to the production process) and inside-out (spillovers from the formerly closed production process) principles to science*" (p. 32).

2.3 Positioning of international organizations

With the rise of interest in Open Science in the highest political forums of the countries, UNESCO started a consultative process[5] on the subject with its member countries, which resulted in the Recommendation on Open Science, adopted during the 41st Session of the UNESCO General Conference in November 2021 [30]. In addition to bringing a consistent definition of Open Science, transcribed below, "*inclusive construct that combines various movements and practices aiming to make multilingual scientific knowledge openly available, accessible and reusable for everyone, to increase scientific collaborations and sharing of information for the benefits of science and society, and to open the processes of scientific knowledge creation, evaluation, and communication to societal actors beyond the traditional scientific community. It comprises all scientific disciplines and aspects of scholarly practices, including basic and applied sciences, natural and social sciences, and the humanities, and it builds on the following key pillars: open scientific knowledge, Open Science infrastructures, science communication, open engagement of societal actors and open dialogue with other knowledge systems*" (p. 7).

The document encourages countries to prioritize seven areas in its implementation and consolidation: "*i. Promoting a common understanding of Open Science, associated benefits, and challenges, as well as diverse paths to Open Science; ii. Developing an enabling policy environment for Open Science; iii. Investing in Open Science*

[5] In 2013, UNESCO held the First Latin American and Caribbean Consultation on Open Access to Scientific Information and Research.

infrastructures and services; iv. investing in human resources, training, education, digital literacy, and capacity building for Open Science; v. fostering a culture of Open Science and aligning incentives for Open Science; vi. Promoting innovative approaches for Open Science at different stages of the scientific process; vii. Promoting international and multi-stakeholder cooperation in the context of Open Science and to reduce digital, technological and knowledge gaps" (p. 6).

Certainly, governments, funding agencies, academic entities, and scientific societies around the world have been promoting actions and policies to stimulate Open Science practices, such as, for example, Horizon Europe, the main driver of Open Science in the European Union, whose strategy is organized into eight action fronts:

- Open data: findable, accessible, interoperable, and reusable (FAIR) data which sharing will be the default in EU-funded scientific research.
- European Open Science Cloud (EOSC): a trusted, virtual, and federated environment that crosses borders and scientific disciplines to store, share, process, and reuse digital research objects[6], following the FAIR principle, and bringing together national and European stakeholders, initiatives, and institutional infrastructures.
- Next-generation metrics: development of new indicators of research quality and impact to complement conventional ones, ensuring fairness to Open Science practices.
- Learning exercise: sharing specific research and innovation challenges in Open Science through the exchange of good practices.
- Future of scientific communication: peer-reviewed scientific publications should be open access and early sharing of research results should be encouraged.
- Rewards: career evaluation systems should fully recognize Open Science activities.
- Research integrity and reproducibility of scientific results: EU-funded research should adhere to commonly agreed standards of research integrity.
- Education and skills: all scientists in Europe should have the necessary skills and support to fully perform with open scientific research.
- Citizen science: society must be able to make meaningful contributions and be recognized as a producer of valid scientific knowledge.

Following this line of action, the program aims, on the one hand, to promote the adoption of Open Science practices, from sharing data and research results (early and widely), stimulating citizen science, and developing new indicators that can be fair in recognizing the commitment to Open Science; on the other hand, to ensure that researchers maintain undisputed intellectual property rights.

---

[6] Publications, data, and digital resources.

2.4 The library and Open Science

2.4.1 Collaboration and multilevel governance

Just as for information science, open access to scientific publications is the cornerstone of Open Science. For librarianship, open research data is one of the most significant contemporary contributions that STL[7] can offer to the community science.

A main definition of open data [31] is "*data that can be freely used, reused, and redistributed by anyone – subject only, at most, to the requirement to attribute and sharealike*" (Guide). Open access to data is the first step of research, its openness can inspire, define, and invigorate research activity. It constitutes the first need of scientific communities and demands from libraries knowledge in efficient and stable storage, data management plans, curation, preservation, formats, and standards, in addition to an understanding of technologies to ensure reuse and sharing [32], to ensure stability, scientific integrity, and collaboration [33]. In the words of Stanton et al. [34]: "*large, collaboratively managed datasets have become essential to many scientific and engineering endeavors, and their management has increased the need for eScience professionals who extend librarianship into solving large-scale information management problems for researchers and engineers*" (p. 79).

Open Science thrives when datasets are accessible and can be shared. Research data management embraces all the product and service characteristics of STL. It is the protagonism conquered "*achieved by taking over the creation and maintenance of institutional repositories. This is a logical consequence of library philosophy that embraces the idea of information for everyone*" ([35], p. 292).

According to Jones [36], the purpose of repositories, in turn, has been extended to include raw research data, their preservation, and reuse, and requires additional investment and skills from libraries and librarians, even though "*many of the data management requirements involve the kind of work in which librarians already have expertise-organizing information, applying metadata standards, and providing access to information*" (Antell [37] apud [33], (p. 7)).

Edwards et al. [38] go further by stating that researchers, when they have time, organize data for themselves or at most in their field of research and not in an interoperable, universal, and shareable way: "*just as with data themselves, creating, handling, and managing metadata products always exacts a cost in time, energy, and attention (...) an additional burden on top of their primary work. Research scientists' main interest, after all, is in using data, not in describing them for the benefit of invisible, unknown future users, to whom they are not accountable and from whom they receive little if any benefit*" (p. 673).

The management of scientific and technological research data is a great opportunity for S&T libraries, also leading them to the preservation and curation activities. The White Paper Envisioning the future of scientific research libraries: a discussion, produced in 2012, advocated the changes needed by libraries and librarians for 21st-century research. How contemporary research information is collected, organized, used, and disseminated, and in particular the movement toward collaboration. Collaboration is particularly timely in addressing the problems of open access to

---

[7] Science and technology libraries are also known as scientific research libraries, research libraries, scientific libraries, university libraries, and academic libraries. Here the term science and technology libraries are adopted.

scientific research data, which deals with concerns common to science and technology. It "*involves a process of joint decision-making among key stakeholders of a problem domain about the future of that domain*" ([39], p. 11) to develop a comprehensive approach to understanding a problem and then acting collectively to solve it.

According to Vangen and Huxham [40], working collaboratively across intra- and inter-organizational boundaries is an indispensable component of organizational life, a way of dealing with issues that cannot be addressed by any one organization acting alone. Historically, libraries have been collaborative, and the future of scholarly research data management requires collaboration at multiple levels and even requires multilevel governance—one of the keys to the survival and success of the academic library in the future. The term multilevel governance and its understanding have spread in various fields of knowledge from political science, international relations, political economy, and public administration. The arrangement connects related and distinct fields around a common interest. The actors involved in multilevel governance consistently act on behalf of the common good and seek a common good as the distinguishing element of multilevel governance from other governance models. These actors will act from political, economic, cultural, social, or scientific coalitions, for example, in coproduction, defined as "*the process through which inputs used to produce a good or service are contributed by individuals who are not in the same organization*" ([41], p. 1073).

From this, it is possible to define multilevel governance [42] "*as a set of general-purpose or functional jurisdictions that enjoy some degree of autonomy within a common governance arrangement and whose actors claim to engage in an enduring interaction in pursuit of a common good*" (p. 4). Multilevel governance in STL is a complex system of governance for the common good, transcends the boundaries of the library in collaborative, multilevel, and multidimensional decision-making processes, involving researchers, information managers, collaborators, partners, and users, in the search for collective solutions for the management and dissemination of information and scientific and technological research data. Effectively, organizational relationships, including relationships in STL, encompass a wide range of arrangements to achieve common goals. At the core of this is a collaboration [43], "*an ongoing negotiation of relationships by individuals who are both participants in the collaboration and, at the same time, accountable to and representative of the diverse organizations and communities involved in, and affected by, it*" (p. 596).

In support of this common good is the philosophical foundation of Open Science. The White Paper [44], which specifically brings out the significant opportunity for library collaboration for Open Science, points out the following actions:

i. Encourage strategic thinking about data sharing;

ii. support the promotion of Open Science standards and policies;

iii. assist in the discovery of open research;

iv. foster partnerships with those acting in Open Science;

v. facilitate collaborations;

vi. participate in inter-institutional projects that leverage open data; and

vii. assist, participate, create, and conduct plans for data management, data warehouses, storage and preservation technologies, and infrastructures such as repositories and data curation.

2.4.2 The Paradigm of Co-Production and Co-Creation.

In the collaborative environment, the term co-creation, introduced in 2004 [45], appears to conceptualize the processes that facilitate the development of a collective approach to a problem, from a situation in which everyone can "*contribute their different perspectives and competencies to the process, and facilitate the joint and shared development of solutions*" (p. 5).

The main objective of the co-creative approach is to facilitate political decision-making through a form of collaborative process design that considers and integrates the widest and most diverse range of relevant perspectives possible [45]. Libraries and librarians are deeply involved in collaborative processes of co-creation of scientific and technological knowledge, fully meeting the description of co-creation and co-production.

Janke and Rush [46] state that librarians (and information professionals) are co-producers of research, as they contribute their expertise throughout the process of knowledge development and dissemination, supporting the quality, improvement, and advancement of research, although "*librarians may not always see their added value, or more strongly, the central role they could play on research teams*" (p. 120). The authors state that: "*every research team needs a librarian who is a core member of the investigative team and not a peripheral member. All too often researchers undervalue and underutilize their expertise because of a lack of awareness of the extent of their skill sets or certain assumptions they have about their contributions to research endeavors*" (p. 118).

2.4.3 The pragmatic wrapping of current scientific information

Engaging in Open Science requires broadening the understanding of access to scientific information. Leading to the consolidation of the term Information Scientist, proposed by Jason Farradane in 1953, on the occasion of the publication of Information service in the industry [47], followed a decade later, in 1961 and 1962 by the conferences of the Georgia Institute of Technology entitled Training Science Information Specialists, founding events of Information Science[8]. The emergent area of Information Science, in the 1960s, was the recognition of a new trend in the scientific field. Its pioneering addresses to the engineer Alexander Ivanovich Mikhaïlov (1905–1988), a member of the founding team of the Russian Institute of Scientific and Technological Information (Viniti) in 1952, he envisaged a career closely linked to the development of scientific and technological information, under the name informatika. He is the main and most influential Russian theorist to address the issue of

[8] Information Science has two roots: on one side Documentation, and the other, Information Retrieval. In the first, what matters is the recording of scientific knowledge, the intellectual memory of civilization, and, in the second, information technologies. Science and Technology were the fertilizing and propulsive elements of its birth, the fruit of the growth of scientific teams, the increase in the number of scientists and researchers, and the acceleration of research, therefore, of knowledge, in addition to technological developments, efforts resulting mainly from World War II. And technologies, especially computers, made it emerge ([48], p. 175).

information production and management in science and technology, an expression used to designate the knowledge obtained from scientific and technological research and development activities. It is an essential input and final product of research and has strategic importance in the development of society.

A significant milestone for Information Science and Technology is the 1963 "*Report Science, government, and information: The responsibilities of the technical community and the government in the transfer of information, known as the Weinberg Report*"[9] [49]. In its summary and major recommendations, it states that to achieve relevance in the results of science and technology, adequate information must be provided for its development. It also emphasizes that *the "transfer of information is an inseparable part of research and development. All those concerned (…) must accept responsibility for the transfer of information in the same degree and spirit that they accept responsibility for research and development itself*" (p. 4). The Weinberg Report urges each public institution engaged in science and technology to accept its responsibility for information activities in areas relevant to the fulfillment of its mission while maintaining its information supply.

Currently, as envisioned in the 1960s and according to the speed of the information society, research is intrinsically linked to the provision of scientific information, especially current scientific information, which can be used in various contexts, from decisions on laboratory funding to access to research data. It is indispensable for the planning and governance of national research, helping to define priorities, allocate funding, and monitor performance. Having the information available to base decisions for scientific and technological development is a capital condition to achieve good results.

However, monitoring the research chain is an ambitious ideal that assumes that the flow of information in the process is organized coherently and reliably, requiring an information system that includes all the steps that contribute to the realization of the research. This system is often described as the Current Research Information System or better known by its acronym CRIS. Interchangeably it may be referred to by the terms Research Information Management System (RIM), Virtual Research Environments, or Research Management Systems, all invoking the idea of information about research management.

The purpose of a CRIS, in effect, is to handle, store, integrate, retrieve, curate, share, and manage information from all phases of scientific research, from inception to publication of its results. It integrates research data from various other systems, to bring together information relevant to the research procedure in a single database, from cooperation opportunities to the dissemination of research results. In general, the information stored in a CRIS includes:

i. Research projects (title, description, duration, academic field, language, whether it is at the institutional or national, or international level, participating institutions, among others);

ii. Researchers (name, affiliations, role in research, area of expertise, educational background, awards, among others);

[9] Alvin M. Weinberg was the chairman of the US President's Science Advisory Committee that presented the report, and so his name became associated with the document.

iii. Organizations involved (name, function or position in research, type of organization, partnerships, among others);

iv. Research input (amount of resources invested in research, time, infrastructure, funding sources, among others);

v. research output (publications, data, patents, awards, products, among others); and

vi. the relationships among all these entities.

Studies on research information management show that standardization, harmonization, and integration of research information often bring great challenges. However, the benefits generated from an integrated dataset of research information are also a major driver of innovation and so Simons [50] points out that CRISs increasingly tend to get a central and fundamental position in this scenario, presenting the idea in **Figures 4** and **5**.

Each element of the ecosystem refers to:

- Funding organizations: distribution of programs, evaluation of results, location of reviewers

- Libraries: acquisition, dissemination

- Researchers: finding collaborations, visibility, profile, reputation, management

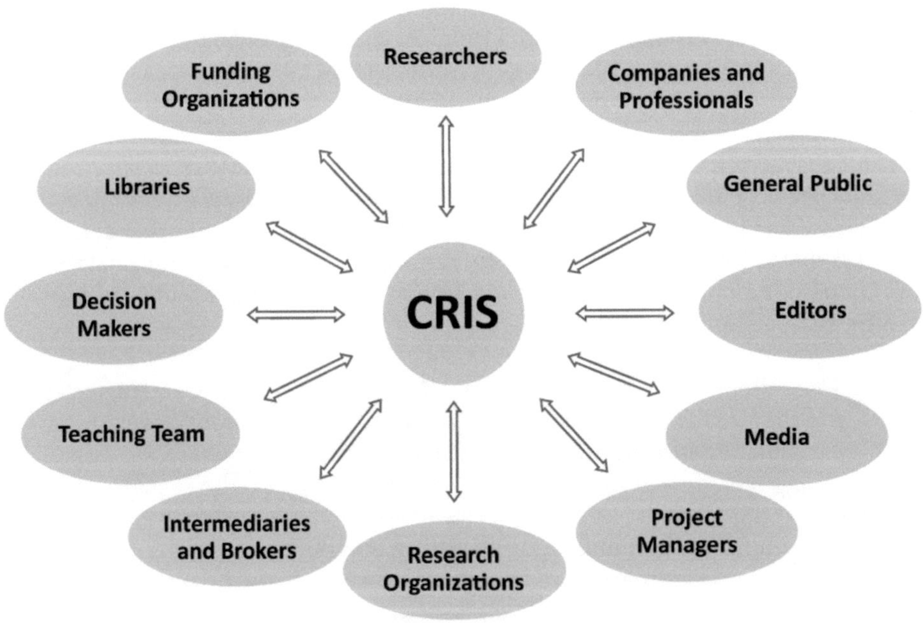

Figure 4.
The centrality of CRIS in the aggregation of information from the scientific and technological research ecosystem (source: Simons [50]).

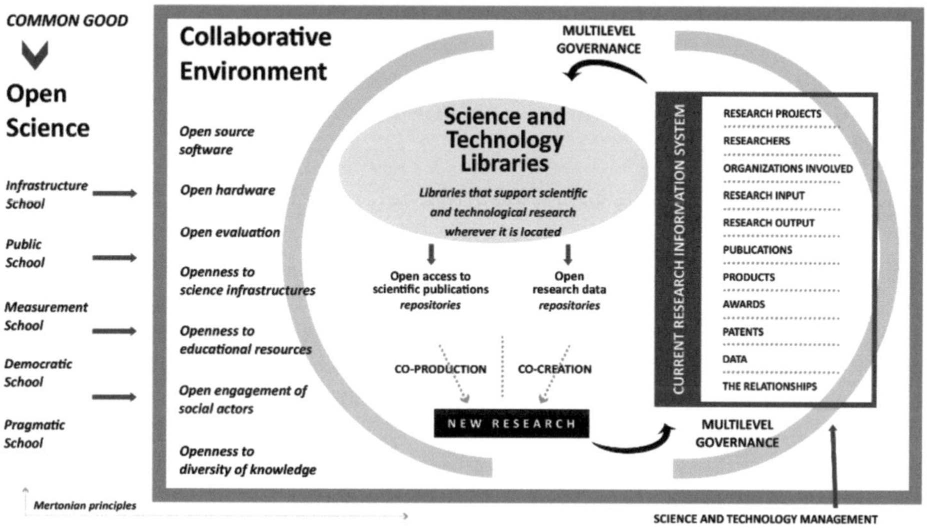

Figure 5.
Framework Open Science collaborative environment in science and technology libraries.

- Decision makers: performance, strategic decisions, priorities, comparisons between countries
- Project managers: overall view, the performance of ongoing activities
- Intermediaries and brokers: finding research results of potential markets or innovation value.
- Teaching team: integration of relevant information into lectures and training
- Research organizations: integration and strategic management of interoperability, profiling
- Editors: finding reviewers
- General public: information and education, interest
- Media: distribution and communication
- Companies and professionals: finding information for participation in projects, partnerships, and use of results.

The underlying principles of CRIS lead compulsorily to the concept of Open Science. Biesenbender et al. [51] point out that, on the one hand, initiatives of open access to scientific information and open data provide links to the CRIS through the interoperability of institutional repositories and data archives. On the other hand, scientists can benefit from CRIS solutions that enable the efficient reuse of information, such as a researcher's record of open-access publications. The CRIS ecosystem, therefore, can be easily perceived as a structuring part of the concept of Open Science.

2.5 Conceptual framework

The concepts presented in this chapter are graphically represented in the framework below, entitled Open Science Collaborative Environment in Science and Technology Libraries.

2.6 Open Science challenges

Open Science is considered one of the main drivers of scientific and technological research. It defends the principles of transparency and accessibility of scientific knowledge, whose main benefits are the strengthening of scientific rigor, increased reliability, acceleration of knowledge dissemination, broader and more inclusive participation of sectors of society in scientific and technological research, effective use of resources and open access to scientific publishing, among others.

However, there are significant challenges to be considered in the adoption of policies in its promotion, among them, the ethical and legal restrictions. In this context are the aspects of personal data protection, intellectual property, and contractual issues. On the other hand, there are the lack of resources, the lack of qualified personnel, the operational issues of the organizations, components that are reflected in the concern to ensure the management, quality and security of shared data, for example.

The complex nature of Open Science, of course, evokes the participation of various stakeholders, in itself a challenging context, which requires promoting the engagement of segments of society both for the co-production of knowledge, analysis, sharing, funding, resources, access to collaborative networks of information and publications in open access, the latter considered the crucial component of Open Science [52], with its own specific challenges. Among them, the same author points out that many researchers who support Open Science do not like the concept of paying to publish, and that open access publishing has given rise to predatory journals, those that publish low-quality articles without adequate peer review in exchange for publication fees. Other problems pointed out by the author are: who should bear the expenses of open access publishing, how to provide equal opportunities for all countries, how to deal with research biases since authors tend to present only positive results, and how open access publication models will evolve.

3. Final considerations

In the conceptual framework that relates S&T library, and Open Science are the three basic rights to use, reuse, and shared knowledge, supported by open digital technologies that guarantee its realization, being, effectively, the transformation of scientific practice. The rise of this ideal must also be represented by the positive consequences of the open predicate: research then becomes more inclusive and more interdisciplinary. Its philosophical basis leads to the understanding that it is necessary to consider Open Science as a possibility to achieve the common good, as advocated by Robert K. Merton's sociology of science. It is achieved through full access to knowledge of science and technology.

In the wake of the philosophy of Open Science are concepts related to open knowledge, such as open-access publication and open data, both with profound significance for librarianship and information science. Science and technology libraries, in particular, regain the leading role in the value chain of scientific and technological research by taking on the creation and maintenance of institutional repositories, expanded to include scientific and technological research data. In dealing effectively with research data, libraries, and librarians need to be prepared to deal with data management, which includes theoretical knowledge of information and computer sciences, for storage, representation, curation, preservation, formats, and standards, to achieve full utilization, reuse, sharing, and wide dissemination.

The discussion, however, does not end with the questions of competence and philosophy. It moves into the political and organizational chapters since Open Science involves intense governance and collaboration. The history of librarianship and the epistemology of information studies show that libraries are collaborative, which is a characteristic (and necessity) of scientific research data management, which requires collaboration across multiple levels and dimensions. Multilevel governance emerges in this scenario for the success of the S&T library of the future, it guides the establishment of political, economic, cultural, social, and scientific coalitions for work in co-production and co-creation, of which librarians should feel as part of the team. Multilevel governance in S&T libraries is a complex system of governance for the common good, it transcends the boundaries of the library in collaborative, multilevel, and multidimensional decision-making processes involving researchers, information managers, collaborators, partners, and users in the search for collective solutions for the management and dissemination of scientific and technological research information and data.

Indeed, co-creative and co-production processes strengthen the viability of joint solutions, since they recognize that no single actor in the process can understand the problem alone and provide a solution. In summary, one can see, in the advancement of Open Science, that the relevance and necessity of librarianship and information science advance concurrently, which in practice is already possible to observe, in the area already deeply engaged in Open Science practices. Librarians and in particular information scientists are mobilized and ready to collaborate in the challenging area of data management, with actions that include selection, storage, organization, representation, validation, and dissemination of data. All actions, to ensure the location, use, sharing, comparison, cross-referencing, and preservation of multidisciplinary data that matter for scientific and technological research. The complexity of the moment for the professions and the institutions involved is in the transition to a world where open access to research will be the standard procedure.

Among the policy options at various levels of the organization to realize multilevel governance in support to Open Science is the pragmatic enveloping of scientific information actions. In particular the monitoring of the current research chain, often described under the terminology current research information system (CRIS), also known as research information management system (RIM) or virtual research environments or research management systems, all invoking the idea of research management information, which in synthesis is information available to make possible the scientific and technological development.

The purpose of a CRIS is the treatment, storage, integration, retrieval, curation, sharing, and management of information in all phases of scientific research, from the beginning to the publication of its results.

Finally, the underlying principles of the CRIS system, compulsorily and reciprocally, lead to the concept of Open Science in several ways. On the one hand, open data and open access to scientific information initiatives provide links to the CRIS through the interoperability of institutional repositories and archives of data. On the other hand, scientists can benefit from CRIS solutions that allow the efficient reuse of information in open access. It is easily perceptive that both are structuring parts of each other.

3.1 Future enhancement

This chapter has dealt with the interplay between S&T libraries, Open Science, organizational collaboration, multilevel governance, and current scientific information. These are complex dimensions of the scientific, policy, and organizational environment, which certainly deserve further exploration and revision. Immediately, it is possible to assess the absence of in-depth discussion about the technological infrastructure to support scientific research, such as for capturing and curating, visualizing and analyzing Open Science data.

Another important gap in this work is the need to deepen the other elements of Open Science (registered in **Figure 2**), such as, for example, the open engagement of social actors, which, like Open Science, bring a new paradigm in the creation of scientific and technological knowledge, with the respective experiential and situational knowledge, equally relevant in the search for true knowledge.

Acknowledgements

The authors would like to thank the Brazilian Institute of Information in Science and Technology (IBICT) for supporting the publication of this work and for sharing the institution's knowledge in the fields of Open Science, Current Research Information System (CRIS) and Science and Technology Libraries (STL).

Conflict of interest

The authors declare no conflict of interest.

Author details

Lillian Alvares* and Kira Tarapanoff
Faculty of Information Science, University of Brasilia, Brasilia, Brazil

*Address all correspondence to: lillianalvares@unb.br

References

[1] Wilson LW. The service of libraries in promoting scholarship and research. Library Quarterly. 1933;**3**:127-145

[2] Buch FB. Bibliothek und geisteswissenschaftliche Forschung. Vandenhoeck and Ruprecht. 1983

[3] Feather J, Sturges P. In: Feather J, Sturges P, editors. International Encyclopedia of Information and Library Science. London: Routledge; 2003

[4] Saunders MNK, Lewis P, Thornhill A. Understanding research philosophy and approaches to theory development. In: Saunders M, Lewis P, Thornhill A, editors. Research Methods for Business Students. 8th ed. Harlow, United Kingdom: Pearson Education Limited; 2019. p. 833

[5] Elo S, Kyngäs H. The qualitative content analysis process. Journal of Advanced Nursing. 2008;**62**(1):107-115

[6] Vercellone C, Bria F, Fumagalli AM, Gentilucci E. Managing the Commons in the Knowledge Economy: Decentralised Citizens Engagement Technologies [Internet]. 2015. Available from: https://www.researchgate.net/publication/282067010

[7] Chubin D. Open science and closed science: Tradeoffs in a democracy. Science, Technology, & Human Values. 1985;**10**(2):73-80

[8] Open Knowledge Foundation. What is open? London

[9] Open Science as a Practice. Was ist Open Science?. Linz

[10] Burgelman JC, Pascu C, Szkuta K, von Schomberg, R. Karalopoulos A, Repanas K, Schouppe M. Open science, open data, and open scholarship: European policies to make science fit for the twenty-first century. Frontiers in Big Data. 2019;2:43

[11] The Netherlands State Secretary for Education Culture and Science. Amsterdam: National Programme Open Science; 2018

[12] Finnish Open Science and Research Initiative. Finland: The Open Science and Research Handbook; 2014

[13] Persic A. Open science diagram from UNESCO. In: Open Science Conference 2021, Leibniz Information Centre for Economics (ZBW). Paris, France: Division of Science Policy and Capacity-Building (SC/PCB), UNESCO; 2021

[14] Suber P. Open Access Overview. Cambridge, Mass: MIT Press; 2012

[15] European Commission. Directorate-General for Research and Innovation. Open Innovation, Open Science, Open to the World: A Vision for Europe. Brussels, Belgium: European Commission; 2016

[16] Federer L, Clarke S, Zaringhalam M, Huerta M. Developing the Librarian Workforce for Data Science and Open Science. 2019

[17] Smith E, Gunashekar S, Lichten C, Chataway J. A framework to monitor Open Science trends in the EU. New Media & Society. 2016;**14**(5):729-747

[18] Pontika N, Knoth P, Cancellieri M, Pearce S. Fostering Open Science to research using a taxonomy and an elearning portal. In: ACM International Conference Proceeding Series.

Association for Computing Machinery; 2015

[19] Nation P, Kyongho H. Where would general service vocabulary stop and special purpose vocabulary begin? System. 1995;**23**(1):35-41

[20] Knoth P, Pontika N. Open Science Taxonomy. 2015

[21] Fecher B, Friesike S. Open science: One term, five schools of thought. In: Opening Science. Cham: Springer International Publishing; 2014. pp. 17-47

[22] Altunay M, Avery P, Blackburn K, Bockelman B, Ernst M, Fraser D, et al. A science driven production cyberinfrastructure: The Open Science grid. Journal of Grid Computing. 2011; **9**(2):201-218

[23] de Roure D, Goble C, Bhagat J, Cruickshank D, Goderis A, Michaelides D, et al. Myexperiment: Defining the social virtual research environment. In: Proceedings of the 4th IEEE International Conference on eScience (eScience 2008). Washington, DC, USA: IEEE Computer Society; 2008. pp. 182-189

[24] Priem J, Taraborelli D, Groth P, Neylon C. Altmetrics: A Manifesto [Internet]. 2010; pp. 1-4. Available from: https://digitalcommons.unl.edu/scholcom/185

[25] Priem J, Groth P, Taraborelli D. The altmetrics collection. PLoS One. 2012;**7**(11)

[26] de Rezende Alvares LMA, de Sá Freire P. Frameworks for Scientific and Technological Research Oriented by Transdisciplinary co-Production. 1st ed. London: Anthem Press; 2022

[27] United Nations General Assembly. Universal Declaration of Human Rights. Paris. New York, USA: United Nations; 1948

[28] West J, Bogers M. Leveraging external sources of innovation: A review of research on open innovation. Journal of Product Innovation Management. 2014;**31**(4):814-831

[29] Tacke O. Open science 2.0: How research and education can benefit from open innovation and web 2.0. In: Bastiaens TJ, Baumöl U, Krämer BJ, editors. On Collective Intelligence. Berlin: Heidelberg, Springer; 2010. pp. 37-48

[30] UNESCO (Organización de las Naciones Unidas para la Educación la C y la C. Proyecto de recomendación sobre la ciencia abierta. In: Proceedings of the 41ª reunión de la Conferencia General. Paris, France: UNESCO; 2021

[31] Open Knowledge Foundation. Open Data Handbook: Guides, Case Studies and Resources for Government & Civil Society on the "What, Why & How" of Open Data. London;

[32] Tzanova S. Changes in academic libraries in the era of Open Science. Education for Information. 2020;**36**(3): 281-299

[33] Schmillen H. Library and Information Science Education and eScience: The Current State of ALA Accredited MLS/MLIS Programs in Preparing Librarians and Information Professionals for eScience Needs [Internet]. 2015. Available from: https://digitalcommons.du.edu/lis_capstone/1

[34] Stanton JM, Kim Y, Oakleaf M, Lankes RD, Gandel P, Cogburn D, et al. Education for eScience professionals: Job analysis, curriculum guidance, and program considerations. Journal of Education for Library and Information Science. 2011;**52**(2):79-94

[35] Maceviciute E. Research libraries in a modern environment. Journal of Documentation. 2014;**70**(2):282-302

[36] Jones C. Institutional Repositories: Content and Culture in an Open Access Environment. Oxford: Chandos Publishing; 2007

[37] Antell K, Foote JB, Turner J, Shults B. Dealing with data: Science librarians' participation in data management at Association of Research Libraries institutions. College & Research Libraries. 2014;**75**(4):557-574

[38] Edwards P, Mayernik S, Batcheller A, Bowker G, Borgman C. Science friction: Data, metadata, and collaboration. Social Studies of Science. 2011;**41**(5):667-690

[39] Gray B. Collaborating: Finding Common Ground for Multiparty Problems. Hoboken, Nova Jersey: Wiley; 1989

[40] Vangen S, Huxham C. Nurturing collaborative relations: Building trust in interorganizational collaboration. The Journal of Applied Behavioral Science. 2003;**39**(1):5-31

[41] Ostrom E. Crossing the great divide: Coproduction, synergy, and development. World Development. 1996;**24**(6):1073-1087

[42] Zürn M, Wälti S, Enderlein H. Introduction. In: Enderlein H, Wälti S, Zürn M, editors. Handbook on Multi-Level Governance. 1st ed. Northampton Massachusetts: Edward Elgar; 2010. p. 504

[43] Lotia N, Hardy C. Critical perspectives on collaboration. In: Cropper S, Ebers M, Huxham C, Ring PS, editors. The Oxford Handbook of Inter-Organizational Relations. 1st ed. Oxford: Oxford University Press; 2009. pp. 595-637

[44] Feltes C, Gibson DS, Miller H, Norton C, Pollock L. Envisioning the Future of Science Libraries at Academic Research Institutions. New York; 2012

[45] Bruhn T, Herberg J, Molinengo G, Oppold D, Stasiak D, Nanz P. Grounded Action Design: A Model of Scientific Support for Processes to Address Complex Challenges. Potsdam; 2019

[46] Janke R, Rush KL. The academic librarian as co-investigator on an interprofessional primary research team: A case study. Health Information and Libraries Journal. 2014;**31**(2):116-122

[47] Farradane JE. Information service in the industry. Research. 1953;**6**(8):327-330

[48] Pinheiro LVR. Ciência da informação, ciências sociais e interdisciplinaridade. Vol. 1. Brasília, Brazil: Instituto Brasileiro de Informação em Ciência e Tecnologia (IBICT); 1999. 182 p

[49] United States President's Science Advisory Committee. Science, Government, and Information: The Responsibilities of the Technical Community and the Government in the Transfer of Information (Weinberg Report) (Report ED 048894). Washington; 1963

[50] Simons E. Introduction to eurocris and cerif. In: 10th Annual Vivo Conference. Podgorica, Montenegro: Radboud University; 2019

[51] Biesenbender S, Petersohn S, Thiedig C. Using current research information systems (CRIS) to showcase national and institutional research (potential): Research information systems in the context of Open Science.

Procedia Computer Science. 2019;**146**: 142-155

[52] Chakravorty N, Sharma CS, Molla KA, Pattanaik JK. Open Science: Challenges, Possible Solutions and the Way Forward. Vol. 88. Proceedings of the Indian National Science Academy. Springer Nature; 2022. pp. 456-471